Guidelines for Developing Spacecraft Maximum Allowable Concentrations for Space Station Contaminants

Subcommittee on Guidelines for Developing Spacecraft Maximum Allowable Concentrations for Space Station Contaminants

DONALD E. GARDNER, *Chairman*, ManTech Environmental Technology, Inc., Research Triangle Park
EULA BINGHAM, University of Cincinnati, Cincinnati
JOSEPH BRADY, Johns Hopkins School of Medicine, Baltimore
RICHARD BULL, Washington State University, Pullman
CHARLES E. FEIGLEY, University of South Carolina, Columbia
MARY ESTHER GAULDEN, University of Texas Southwestern Medical School, Dallas
PAUL F. HALFPENNY, Engineering and Consulting, Van Nuys, CA
WILLIAM E. HALPERIN, National Institute for Occupational Safety and Health, Cincinnati
ROGENE F. HENDERSON, Lovelace Biomedical and Environmental Research Institute, Albuquerque
MICHAEL HOLICK, Boston University School of Medicine, Boston
E. MARSHALL JOHNSON, Thomas Jefferson Medical College, Philadelphia
RALPH L. KODELL, National Center for Toxicological Research, Jefferson, AR
DANIEL KREWSKI, Health and Welfare Canada, Ottawa, Ontario
ROBERT SNYDER, Rutgers University, Piscataway
KATHLEEN TAYLOR, General Motors Research Laboratories, Warren, MI
BERNARD M. WAGNER, Wagner Associates, Inc., Millburn, NJ
G. DONALD WHEDON, Shriners Hospitals for Crippled Children, Tampa

Staff

KULBIR S. BAKSHI, Project Director
RICHARD D. THOMAS, COT Program Director
MARVIN A. SCHNEIDERMAN, Senior Staff Scientist
RUTH E. CROSSGROVE, Editor
BEULAH S. BRESLER, Senior Editorial Assistant
CATHERINE M. KUBIK, Senior Program Assistant

Committee on Toxicology

JOHN DOULL, *Chairman*, University of Kansas Medical Center, Kansas City
EULA BINGHAM, *Vice-Chairman*, University of Cincinnati, Cincinnati
R. HAYS BELL, Eastman Kodak Company, Rochester, NY
DEAN E. CARTER, University of Arizona, Tucson
CHARLES E. FEIGLEY, University of South Carolina, Columbia
DONALD E. GARDNER, ManTech Environmental Technology, Inc., Research Triangle Park
MARY ESTHER GAULDEN, University of Texas Southwestern Medical School, Dallas
WALDERICO GENEROSO, Oak Ridge National Laboratory, Oak Ridge
IAN GREAVES, University of Minnesota, Minneapolis
RONALD A. HITES, Indiana University, Bloomington
CAROLE A. KIMMEL, Environmental Protection Agency, Washington, DC
RALPH L. KODELL, Food and Drug Administration, National Center for Toxicological Research, Jefferson, AR
LOREN D. KOLLER, Oregon State University, Corvallis
ERNEST EUGENE MCCONNELL, Raleigh, NC
ROBERT SNYDER, Rutgers University, Piscataway
BERNARD M. WAGNER, Wagner Associates, Inc., Millburn, NJ
BAILUS WALKER, JR., University of Oklahoma, Oklahoma City
HANSPETER R. WITSCHI, University of California, Davis
GAROLD S. YOST, University of Utah, Salt Lake City

Staff

RICHARD D. THOMAS, Program Director
KULBIR S. BAKSHI, Senior Program Officer
MARVIN A. SCHNEIDERMAN, Senior Staff Scientist
BEULAH S. BRESLER, Senior Editorial Assistant
CATHERINE M. KUBIK, Senior Program Assistant

Board on Environmental Studies and Toxicology

PAUL G. RISSER, *Chairman*, University of New Mexico, Albuquerque
FREDERICK R. ANDERSON, Cadwalader, Wickersham & Taft, Washington, DC
JOHN C. BAILAR III, McGill University, Faculty of Medicine, Montreal
LAWRENCE W. BARNTHOUSE, Oak Ridge National Laboratory, Oak Ridge
GARRY D. BREWER, University of Michigan, Ann Arbor
EDWIN H. CLARK, Department of Natural Resources & Environmental Control, State of Delaware, Dover
YORAM COHEN, University of California, Los Angeles
JOHN L. EMMERSON, Lilly Research Laboratories, Greenfield, IN
ROBERT L. HARNESS, Monsanto Agricultural Company, St. Louis
ALFRED G. KNUDSON, Fox Chase Cancer Center, Philadelphia
GENE E. LIKENS, The New York Botanical Garden, Millbrook
PAUL J. LIOY, UMDNJ-Robert Wood Johnson Medical School, Piscataway
JANE LUBCHENCO, Oregon State University, Corvallis
DONALD MATTISON, University of Pittsburgh, Pittsburgh
GORDON ORIANS, University of Washington, Seattle
NATHANIEL REED, Hobe Sound, FL
MARGARET M. SEMINARIO, AFL/CIO, Washington, DC
I. GLENN SIPES, University of Arizona, Tucson
WALTER J. WEBER, JR., University of Michigan, Ann Arbor

Staff

JAMES J. REISA, Director
DAVID J. POLICANSKY, Associate Director and Program Director for Natural Resources and Applied Ecology
RICHARD D. THOMAS, Associate Director and Program Director for Human Toxicology and Risk Assessment
LEE R. PAULSON, Program Director for Information Systems and Statistics
RAYMOND A. WASSEL, Program Director for Environmental Sciences and Engineering

Commission on Life Sciences

BRUCE M. ALBERTS, *Chairman*, University of California, San Francisco
BRUCE N. AMES, University of California, Berkeley
J. MICHAEL BISHOP, University of California Medical Center, San Francisco
MICHAEL T. CLEGG, University of California, Riverside
GLENN A. CROSBY, Washington State University, Pullman
LEROY E. HOOD, California Institute of Technology, Pasadena
DONALD F. HORNIG, Harvard School of Public Health, Boston
MARIAN E. KOSHLAND, University of California, Berkeley
RICHARD E. LENSKI, Michigan State University, East Lansing
STEVEN P. PAKES, University of Texas, Dallas
EMIL A. PFITZER, Hoffmann-LaRoche, Inc., Nutley, NJ
THOMAS D. POLLARD, Johns Hopkins Medical School, Baltimore
JOSEPH E. RALL, National Institutes of Health, Bethesda, MD
RICHARD D. REMINGTON, University of Iowa, Iowa City
PAUL G. RISSER, University of New Mexico, Albuquerque
HAROLD M. SCHMECK, JR., Armonk, NY
RICHARD B. SETLOW, Brookhaven National Laboratory, Upton, NY
CARLA J. SHATZ, University of California, Berkeley
TORSTEN N. WIESEL, Rockefeller University, New York

JOHN E. BURRIS, Executive Director

Acknowledgments

The Committee on Toxicology's Subcommittee on Spacecraft Maximum Allowable Concentrations for Space Station Contaminants gratefully acknowledges the valuable assistance provided by the following personnel from the National Aeronautics and Space Administration (NASA) and its contractors:

Dr. John James (NASA)
Dr. Duane Pierson (NASA)
Dr. Martin Coleman (NASA)
Dr. Lawrence Dietlein (NASA)
Mr. Jay Perry (NASA)
Mr. Kenneth Mitchell (NASA)
Mr. James Hyde (Jet Propulsion Laboratory)
Dr. King Lit Wong (Krug International)
Mr. Donald Cameron (Boeing Company)

The subcommittee also acknowledges the valuable assistance provided by the Johnson Space Center, Houston, Texas, the Marshall Space Flight Center, Huntsville, Alabama, the Kennedy Space Center, Cape Canaveral, Florida, and the Space Station Freedom Program Office, Reston, Virginia, for providing tours of their facilities. The subcommittee is grateful to astronauts Dr. Shannon Lucid and Dr. Drew Gaffney for sharing their experiences.

Preface

The National Aeronautics and Space Administration (NASA) has been concerned with the potential toxicological hazards to humans that may be associated with entering and living within spacecraft for prolonged periods. Despite major engineering advances in controlling the atmosphere within spacecraft, some contamination of the air appears inevitable. NASA has responded to this problem by measuring and identifying numerous airborne contaminants during each space mission. As space missions increase in duration and complexity, ensuring the health and well-being of individuals traveling and working in this unique environment becomes increasingly uncertain.

As part of its attempt to establish a safe space environment, NASA has requested that the National Research Council's Committee on Toxicology (COT) develop appropriate guidelines for establishing spacecraft maximum allowable concentrations (SMACs) for space-station contaminants. SMACs are not intended to represent the absolute dividing line between safe and unsafe exposures, because certain individuals may respond differently. Short-term SMACs provide guidance for crew members when it becomes necessary to operate under emergency conditions lasting up to 24 hr. Such exposures may produce measurable, but reversible, responses and discomfort, but they should not interfere with or impair judgment. Long-term SMACs are intended to avoid adverse health effects, immediate and delayed, and to avoid degradation in performance from continuous exposure.

In response to NASA's request, COT established the Subcommittee on Guidelines for Developing Spacecraft Maximum Allowable Concentrations for Space Station Contaminants. This subcommittee was made up of scientists with expertise in toxicology, biochemistry, epidemiology, pathology, analytical chemistry, industrial hygiene, genetics, occupational medicine, biostatistics, and risk assessment.

In establishing these guidelines, several major issues were considered, including the development of methods for (1) translating existing animal toxicity data to predict toxicity for humans in space, (2) evaluating the toxicity of complex mixtures, (3) modifying risk estimates to account for altered physiological changes and stresses caused by the microgravity environment of space, and (4) determining how other exposure conditions within the spacecraft, such as recirculation of air and water, lack of purging of expired air, activities outside the space station, contamination from airlocks, reentry, and radiation, may alter the predicted or expected response.

The subcommittee recognizes that many factors, such as the alterations in normal human physiological and biochemical processes associated with spaceflight, are not fully understood and could warrant revisions of proposed SMAC values. The subcommittee emphasizes the necessity to review continuously and adjust these values when additional scientific data become available.

This report could not have been produced without the untiring efforts of the National Research Council staff: Beulah Bresler, senior editorial assistant; Catherine Kubik, senior program assistant; and Ruth E. Crossgrove, editor.

Finally, the subcommittee gratefully acknowledges the persistence, patience, and expertise of Dr. Kulbir S. Bakshi, project director of the subcommittee, and Dr. Richard D. Thomas, program director of COT, in bringing this report to its final form.

Donald E. Gardner, *Chairman*
Subcommittee on Guidelines for Developing
 Spacecraft Maximum Allowable Concentrations
 for Space Station Contaminants

John Doull, *Chairman,*
Committee on Toxicology

Contents

LIST OF ABBREVIATIONS	xiii
EXECUTIVE SUMMARY	1
1 INTRODUCTION	9
2 HISTORICAL DEVELOPMENT OF LIMITS FOR TOXICANTS IN SPACECRAFT	13
3 SOURCES OF SPACE-STATION CONTAMINANTS	23
4 ENVIRONMENTAL CONTROL AND LIFE-SUPPORT SYSTEM	41
Contaminant Control and Monitoring	42
Recommendations	44
5 PHYSIOLOGICAL CHANGES DURING SPACEFLIGHT	49
Space Motion Sickness	49
Bone and Mineral Metabolism	50
Muscle Metabolism	53
Cardiovascular Function and Body Fluid Changes	54
Immune System	55
Nutrition	55
6 ESTABLISHMENT OF SPACECRAFT MAXIMUM ALLOWABLE CONCENTRATIONS	59
Description of SMACs	59
Sources of Data for Developing SMACs	60
Types of Data Used in Recommending SMACs	64
Risk Assessment	78
General Approach to Establishing SMACs	94

REFERENCES 97

APPENDIX 1
Format for SMAC Documents 109

APPENDIX 2
Conversion Factors 111

APPENDIX 3
Reference Values Used by COT 113

List of Abbreviations

ACGIH   American Conference of Governmental Industrial Hygienists
ACMA   Atmospheric composition monitor assembly
ACS   Atmosphere control and supply
AD   Aerodynamic diameter
ADI   Acceptable daily intake
ALARA   As low as reasonably achievable
ARS   Atmosphere revitalization subsystem
ATSDR   Agency for Toxic Substances and Disease Registry
BEST   Board on Environmental Studies and Toxicology
CEGL   Continuous exposure guidance level
CNS   Central nervous system
COT   Committee on Toxicology
ECLSS   Environmental control and life-support system
EEGL   Emergency exposure guidance level
EEL   Emergency exposure limit
EPA   Environmental Protection Agency
FAA   Federal Aviation Administration
FDS   Fire detection and suppression
IRIS   Integrated Risk Information System
LOAEL   Lowest-observed-adverse-effect level
MCA   Major constituent analyzer
MF   Modifying factor
NASA   National Aeronautics and Space Administration
NIOSH   National Institute for Occupational Safety and Health
NOAEL   No-observed-adverse-effect level
NOEL   No-observed-effect level
NRC   National Research Council
OSHA   Occupational Safety and Health Administration
PEL   Permissible exposure limit
PMC   Permanently manned capability
RDA   Recommended Dietary Allowance

REL	Recommended exposure limit
RfC	Reference concentration
RfD	Reference dose
SF	Safety factor
SMAC	Spacecraft maximum allowable concentration
SMS	Space motion sickness
STEL	Short-term exposure limit
TCCS	Trace contaminant control subassembly
THC	Temperature and humidity control
TLV	Threshold limit value
UF	Uncertainty factor
WM	Waste management
WRM	Water recovery and management

Executive Summary

The National Aeronautics and Space Administration (NASA) is preparing to launch a manned space station—Space Station Freedom—by the mid-1990s. Because Space Station Freedom will be a closed complex environment, some contamination of its atmosphere is inevitable. Several hundred chemicals are likely to be found in the closed atmosphere of the space station, most in very low concentrations. Important sources of atmospheric contaminants include metabolic waste products of crew members and off-gassing of cabin materials and equipment. Release of chemicals from experiments performed on board the space station is also a possible source of contamination, and the water reclamation system has the potential to introduce novel compounds into the air. NASA is concerned about the health, safety, and functional abilities of crews exposed to these contaminants.

This report, prepared by the Committee on Toxicology of the National Research Council's Board on Environmental Studies and Toxicology, is in response to a request from NASA for guidelines to develop spacecraft maximum allowable concentrations (SMACs) for space-station contaminants. SMACs are used to provide guidance on allowable chemical exposures during normal operations and emergency situations. Short-term SMACs refer to concentrations of airborne substances (such as gas, vapor, or aerosol) that will not compromise the performance of specific tasks during emergency conditions lasting up to 24 hr. Exposure to 1- or 24-hr SMACs will not cause serious or permanent effects but may cause reversible effects that do not impair judgment or interfere with proper responses to emergencies such as fires or accidental releases.

Long-term SMACs are intended to avoid adverse health effects (either immediate or delayed) and to avoid degradation in crew performance with continuous exposure in a closed space-station environment for as long as 180 days. Chemical accumulation, detoxification, excretion, and repair of toxic insults are thus important in determining 180-day SMACs.

ENVIRONMENTAL CONTROL AND LIFE-SUPPORT SYSTEM

The environmental control and life-support system (ECLSS) of the space station is designed to control temperature, humidity, and composition of space-station air, including CO_2 removal; recover water; dispose of waste; and detect and suppress fires. Fires are a great potential hazard and much attention has been given to suppressing them. A fire suppression system is available, but if all else fails, an escape vehicle can be used. A subsystem of the ECLSS, the atmosphere revitalization system, which includes a mass spectrometer called the major constituent analyzer, will analyze cabin air for O_2, N_2, H_2, CO, H_2O, and CH_4 in all areas of the habitation and laboratory modules. A design criterion for the atmosphere revitalization subsystem is the maintenance of space-station exposure levels below the 180-day SMACs under normal conditions.

MODIFICATION OF CONTAMINANT TOXICITY BY ENVIRONMENTAL FACTORS

The special conditions of the space environment must be taken into account in defining spacecraft contaminant exposure limits. Deposition of particles is clearly different and lung function and the toxic potential of inhaled particles may be different under microgravity conditions than under full gravity conditions, as on earth.

Astronauts will be physically, physiologically, and psychologically compromised for the following reasons: loss of muscle and bone mass, altered immune system, cardiovascular changes, decreased red-blood-cell mass, altered nutritional requirements, behavioral changes from stress, fluid shift in the body, altered hormonal status, and altered drug metabolism. These changes could be important factors in disease susceptibility.

The physiological changes noted in spaceflight to date demonstrate that the astronaut is in an altered homeostatic state and thus may be more susceptible to toxic chemicals. How this altered state modifies reactions to chemicals in the space-station environment is not fully known. The physiological changes induced in the space crew are important and their impact must be taken into account in developing SMAC values for various contaminants.

SOURCES AND TYPES OF DATA FOR ESTABLISHMENT OF SMACs

The subcommittee recommends the use of data derived from a number of sources in establishing SMAC values. These sources provide information on a variety of health effects, including mortality, morbidity, clinical signs and symptoms, pulmonary effects, neurobehavioral effects, immunotoxicity, reproductive and developmental toxicity, pathology, mutagenicity, carcinogenicity, and biochemical and enzyme changes.

Chemical-Physical Characteristics of Toxicants

The chemical and physical characteristics of a substance provide valuable information on potential tissue dosimetry of the compound within the body and on its likely toxic effects. Preliminary estimates of the toxic potential of new chemicals also may be derived from known toxicities of structurally similar, well-investigated compounds. However, additional uncertainty (safety) factors must be applied to arrive at safe levels for those congeners that have no dose-response data from intact animals.

In Vitro Toxicity Studies

Useful information can be obtained from studies conducted to investigate adverse effects of chemicals on cellular or subcellular systems in vitro. Systems in which toxicity data have been collected include isolated organ systems, single-cell systems, and tissue cultures from multicellular organisms maintained under defined conditions or from functional units derived from whole cells. In vitro studies can be used to elucidate the toxic effects of chemicals and to study their mechanism of action.

Animal Toxicity Studies

The data necessary to evaluate the relationship between exposure to a toxic chemical and its effects on people are frequently not available from human experience; therefore, animal toxicity studies must be relied on to provide information on responses likely to occur in humans.

The usefulness of animal data depends in part on the route of exposure and species used. Although inhalation studies are most relevant in assessing the toxicity of atmospheric contaminants, data from skin absorption, ingestion, and parenteral studies are also useful. The relevance of animal data to humans may be limited by the absence of information on affected target organs and knowledge of metabolic pathways and pharmacokinetics in animals and humans.

Clinical and Epidemiological Observations

In establishing SMACs for chemicals, dose-response data from human exposure should be used whenever possible. Data from clinical inhalation exposures are most useful because inhalation is the most likely route of exposure. Human toxicity data also are available from epidemiological studies of long-term industrial exposures, from short-term high-level exposures following accidents, or from therapeutic uses of some pharmaceutical agents. Some of these data provide a basis for estimating a dose-response relationship.

Epidemiological studies have contributed to our knowledge of the health effects of many airborne chemical hazards. The limitations of epidemiology stem from its use of available data. The accuracy of data on health outcomes varies with the source of the information, and records documenting historical exposure levels are often sparse. For example, mortality information derived from death certificates is sometimes inaccurate, and exposure information collected for administrative purposes is limited. Despite these limitations, if the populations studied are large enough and have been exposed to high enough doses over a sufficient period to allow for the expression of disease, epidemiological studies usually provide valuable information on the effects of exposure in humans without resorting to cross-species extrapolation or to exposing humans in an experimental situation to possible injuries from chemical hazards.

Pharmacokinetics and Metabolism

Evaluation of the health effects of any chemical in a given environment is greatly facilitated by an understanding of its physiological disposition in the body. Many chemicals require some form of metabolic activation to exert their toxic effects. The formation of reactive metabolites may depend on the level of exposure and the pharmacokinetics of the chemical. Modern pharmacokinetic studies can provide physiologically based models describing disposition of chemicals within organs and tissues in the body. The space station is a closed system with limited capacity to clear the air of chemical vapors; the crew contributes to the removal of the chemicals from the air through sequestration and metabolism.

Toxic metabolites may be highly reactive chemically. These metabolites are biologically reactive intermediates that may covalently bind to nucleic acids or proteins that, in turn, may alter DNA replication or transcription. In addition to formation of reactive metabolites, metabolic activity also may lead to formation of species of active oxygen that may damage nucleic acids or proteins or cause lipid peroxidation. The resulting health effects may range from direct, short-term target-organ toxicity to carcinogenesis.

Biological Markers

Biological markers are indicators of change within an organism that link exposure to a chemical to subsequent development of adverse health effects. Biological markers within an exposed individual can indicate the degree of exposure to a pollutant and may provide evidence of the initial structural, functional, or biochemical changes induced by the exposure and, ultimately, the biochemical or physiological changes associated with adverse health effects.

Biological markers can be divided into three classes:

1. Biological markers of exposure to pollutants may be thought of as "footprints" that the chemical leaves behind upon interaction with the body. Such markers contain the chemical itself or a metabolic fragment of the chemical and thus are usually chemical-specific.

2. Biological markers of the effects of exposure include the totality of subclinical and clinical signs of chemically induced disease states. The markers of greatest interest are those that are early predictors of

serious effects or late-occurring effects. Such markers would be useful in determining what levels of pollutants in the space station can be tolerated without causing irreversible deleterious health effects.

3. Biological markers of increased susceptibility to the effects of exposure to pollutants could be used to predict which persons are most likely to be at excess risk as space-station crew members.

RISK ASSESSMENT
(DEVELOPMENT OF EXPOSURE CRITERIA)

The assessment of toxicants that do not induce carcinogenic or mutagenic effects traditionally has been based on the concept that an adverse health effect will not occur below a certain level of exposure, even if exposure continues over a lifetime. Given this assumption, a reference dose intended to avoid toxic effects may be established by dividing the no-observed-adverse-effect level by an appropriate uncertainty factor or set of factors.

For carcinogenic effects, especially those known to be due to direct mutagenic events, no threshold dose may exist. However, when carcinogenesis is due to epigenetic or nongenotoxic mechanisms, a threshold dose may be considered. Attempts to estimate carcinogenic risks associated with levels of exposure have involved fitting mathematical models to experimental data and extrapolating from these models to predict risks at doses that are usually well below the experimental range. The multistage model of Armitage and Doll is used most frequently for low-dose extrapolation. According to multistage theory, a malignant cancer cell develops from a single stem cell as a result of a number of biological events (e.g., mutations) that must occur in a specific order. Recently, a two-stage model that explicitly provides for tissue growth and cell kinetics also has been used in carcinogenic risk assessment.

The multistage model, characterized by low-dose linearity, forms the basis for setting SMACs for carcinogens. Low-dose linearity is generally assumed for chemical carcinogens that act through direct interaction with genetic material.

ISSUES IN MAKING RECOMMENDATIONS FOR THE ESTABLISHMENT OF SMACs

A number of issues need to be considered in developing recommendations for establishing SMACs. These issues include (1) translating animal toxicity data to predict toxicities in humans; (2) determining 30- or 180-day SMACs for carcinogens based on toxicological or epidemiological studies that often involve long-term or lifetime exposure; (3) considering limits set by the Occupational Safety and Health Administration, the American Conference of Governmental Industrial Hygienists, and the National Research Council in developing SMACs; (4) evaluating the toxicities of mixtures; and (5) modifying risk assessments based on the altered environment in the microgravity of space.

1 Introduction

The National Aeronautics and Space Administration (NASA) is preparing to launch a manned space station by the year 1996. Because of concerns about the health, safety, and functioning abilities of the crews, NASA has requested that the National Research Council (NRC) through the Board on Environmental Studies and Toxicology (BEST) provide advice on toxicological matters for the space-station program. The Subcommittee on Guidelines for Developing Spacecraft Maximum Allowable Concentrations for Space Station Contaminants was established by the Committee on Toxicology (COT) to address NASA's concerns.

Spacecraft maximum allowable concentrations (SMACs) are defined as the maximum concentrations of airborne substances (such as gas, vapor, or aerosol) that will not cause adverse health effects, significant discomfort, or degradation in crew performance. SMACs are further classified into 1- or 24-hr emergency SMACs and 7-, 30-, or 180-day continuous SMACs. The 1- or 24-hr SMACs are to be used in emergency situations such as accidental spills or fire. The 7-, 30-, and 180-day SMACs are guidance levels intended to avoid adverse health effects, either immediate or delayed, and to avoid degradation in performance of crew after continuous exposure for as long as 180 days in the space station; these values will be used for normal operations of the space station. NASA expects that each crew for the space station will stay aloft for up to 180 days. Limits for continuous exposure to toxic substances for up to 1,000 days have been recommended for a few chemicals by NRC (1968, 1972). However, the rationale for these limits was not discussed in these reports. More recently, COT has recommended exposure guidance levels for about 70 chemicals for up to 90 days for crews of submarines (NRC, 1984a,b,c, 1985a,b, 1986a, 1987a, 1988a) with extensive documentation and rationale. In setting 7-, 30-, and 180-day SMACs for the space station, the subcommittee will have to consider not only a longer exposure period but also the physiological changes induced by the spaceflight and the susceptibility of the crew to the effects of these changes. The subcommittee will also consider the question of exposure to mixtures of chemicals (approximately 300 substances have been identified that are expected to be found in the space station).

The current space-station program began with President Reagan's 1984 State of the Union address in which he directed NASA to develop a permanently manned space station within a decade. The space station that NASA is now planning in response to this directive will be 353 feet long. It will be resuppliable and capable of staying in orbit for about 30 years. Many of its systems will be highly automated, leaving the crew more time for the tasks that can be performed only by humans. The space station will be a multipurpose national research facility with the following functions:

- A laboratory in which long-term research and technology development studies can be conducted in the almost perfect vacuum and microgravity of space.
- A manufacturing facility to make pure pharmaceuticals, perfect crystals, and exotic metal alloys.
- A laboratory for life sciences research conducted in microgravity.
- A permanent observatory for earth or stellar viewing in low earth orbit.

The space station will be continuously habitable; it will depend on the space shuttle for launch and assembly as well as for crew rotation. As currently planned, it will take nearly 20 space-shuttle flights to assemble and outfit the space station in orbit. Initial plans call for a crew of four persons in the permanently manned phase of the operational space station, supported by resupply and crew-rotation flights of the space shuttle every 3 months. Wash water and condensate will be purified and reused. Present planning calls for food and nitrogen to be periodically resupplied from earth. Other materials to be brought to the space station include oxygen to replace gradual loss through leakage, raw materials for materials processing, and propellants for engines needed to correct orbital decay; miscellaneous supplies also will be brought up every few months.

The crew of the space station will live and operate primarily in four (or more) pressurized modules. The U.S. laboratory and habitation modules will be approximately 27 feet long and 15 feet in diameter. The modules will be interconnected by resource nodes and various support structures. At PMC (permanently manned capability) the pressure level inside will be the same as on earth (14.7 psi). The space station will operate at an altitude between 150 and 270 nautical miles.

The space-station program involves international participation. NASA has signed memoranda of understandings with Canada, the European Space Agency, and Japan. These countries and space organizations will work with NASA to design and build some components of the space station. Personnel from these countries also will be included in the space-station crews. The Space Station Freedom design currently comprises one module for habitation and three modules for laboratories. All four of the modules will be interconnected.

The space station will be an essentially closed and very complex environment; therefore, some contamination of its atmosphere is inevitable. Several hundred chemicals have been found in samples of the atmosphere from space-shuttle missions. Even more chemicals are likely to be found in the space station. The contaminants come from humans, from materials making up the spacecraft, from operation of equipment, from the atmospheric gas supply, from food, from supplies, from experiments or manufacturing activities on the space station, and from reactions and interactions that occur between ordinarily nontoxic constituents within a closed space. As manned space missions increase in duration, so will the severity and complexity of the contamination problem and the necessity for control and removal of these contaminants. Before an appropriate trace contaminant control system can be designed, the generation of atmospheric contaminants must be understood and their allowable exposure levels must be properly defined.

In response to NASA's request, this report was prepared to set forth guidelines for establishing SMACs for up to 180 days for potential atmospheric contaminants in the closed environments of the space station. Where appropriate knowledge exists, these guidelines take into account the pronounced physiological changes and stresses that occur in microgravity environments during spaceflight and the behavior of the contaminants under these conditions.

Some of the issues that should be addressed in developing these guidelines include the following:

- Converting animal or human toxicity data derived from inhalation, oral, or parenteral exposures to human 1-hr, 24-hr, 7-day, 30-day, and 180-day SMACs.

12 GUIDELINES FOR DEVELOPING SMACS

- Examining the issue of exposure to mixtures of chemicals (simultaneous exposure to approximately 300 chemicals) and proposing maximum exposure levels to these mixtures for the crew.
- Modifying the current methods for developing appropriate exposure criteria by taking into account the pronounced physiological changes and stresses that occur in the microgravity of space.

2 Historical Development of Limits for Toxicants in Spacecraft

In 1968, at the request of NASA, the NRC's Space Science Board organized the Panel on Air Standards for Manned Space Flight to examine the likelihood of adverse effects from air contaminants on the health or performance of spacecraft crews on prolonged missions and to recommend limits for human exposure to spacecraft contaminants under emergency conditions. Approximately 200 potential contaminants of spacecraft atmosphere, identified from simulated spacecraft studies and off-gassing experiments, were reviewed. Some of these substances had already been examined by COT, and exposure limits had been recommended for submarine atmospheres. The Space Science Board panel recommended the acceptance of these limits for 90-day missions. From the 200 possible contaminants, the panel identified compounds having significant hazard potential for which no submarine standards had been recommended and asked NASA to set priorities for chemicals of primary concern. NASA identified 11 compounds for which they desired long-term exposure limits; NASA also requested emergency limits for five compounds.

The recommended exposure limits were based on exhaustive analyses of the toxicity literature on these compounds. The 90-day long-term limits recommended were chosen with the objective of avoiding (1) adverse health effects, either immediate or delayed, and (2) degradation of performance. The 1-hr emergency limits were designed to avoid significant degradation in crew performance in emergencies and to avoid permanent injury; no safety factors were included, and transitory effects were considered acceptable.

The recommendations of this panel included (1) acceptance of the 90-day continuous exposure limits recommended by COT for 23 contaminants for 90-day submarine missions (Table 1); (2) adoption of limits, as provisional guidelines for spaceflight, for 11 contaminants for 90-day and 1,000-day missions (Table 2); and (3) adoption of 60-min provisional emergency exposure limits (EELs) for five contaminants in the event of a single accidental exposure during the mission (Table 3) (NRC, 1968). It should be noted that many of the values in Tables 1-3 may be out of date because new toxicological data have become available since they were originally developed.

TABLE 1 90-Day Contaminant Concentration Limits[a]

Contaminant	90-Day Limit, mg/m^3
Acetone	71
Acetylene	2,700
Ammonia	17
Benzene	3
Carbon monoxide	29
Chlorine	0.3
Dichlorodifluoromethane	5,000
1,1,2,2-Tetrafluoro-1,2-dichloroethane	7,000
Ethyl alcohol	115
Hydrogen	245
Hydrogen chloride	1.5
Hydrogen fluoride	0.1
Methane	3,300
Methyl alcohol	13
Methyl chloroform	3,000
Monoethanolamine	1
Nitrogen dioxide	1
Ozone	0.04
Phosgene	0.2
Sulfur dioxide	2.6
Toluene	188
1,1,1-Trichloroethane	1,100
Xylene	217

[a]Adapted from NRC (1968).

In 1971, NASA requested that COT review the exposure limits established in 1968 by the Space Science Board (NRC, 1968) and set new limits where appropriate. The new limits were of primary importance to NASA in connection with providing engineering benchmarks to

TABLE 2 Provisional Limits for Space-Cabin Contaminants for 90 and 1,000 Days[a]

Contaminant	Air Limit in mg/m^3	
	90 Days	1,000 Days
n-Butanol	30	30
2-Butanone	58	59
Carbon monoxide	17	17
Chloroform	24	5
Dichloromethane	105	21
Dioxane	36	7
Ethyl acetate	144	144
Formaldehyde	0.12	0.12
2-Methylbutanone	82	82
Trichloroethylene	54	11
1,1,2-Trichloro, 1,2,2-trifluoroethane and related congeners	161	No recommendation

[a]Adapted from NRC (1968).

guide the development of advanced life-support systems for long-term missions. COT established the Panel on Air Quality in Manned Spacecraft for this review, and provisional limits were recommended for 52 potential spacecraft contaminants for a variety of exposure durations (Table 4) (NRC, 1972). The panel emphasized that each limit value was provisional and subject to change as additional data became available. The panel also emphasized that these limits represented a maximum allowable concentration of a single contaminant without regard to its occurrence in mixtures of contaminants. For toxicological assessment of contaminant mixtures, the panel recommended the use of a group-limit concept (for greater detail, see pages 91-92).

TABLE 3 Provisional Emergency Limits for Space-Cabin Contaminants[a,b]

Contaminant	Air Limit in mg/m^3 for 60 min
2-Butanone	294
Carbonyl fluoride	67
Ethylene glycol	253
2-Methylbutanone	409
1,1,2-Trichloro, 1,2,2-trifluoroethane and related congeners	1,612

[a]Adapted from NRC (1968).
[b]Applies to a single exposure during the mission.

TABLE 4 Maximum Allowable Concentrations for Manned Spacecraft[a]

Compound (Molecular Weight)	ppm (mg/m^3)				Notes
	10 Min, Special Area	60 Min	90 Days	6 Mon	
ALCOHOLS					
Methyl alcohol (32.04)	—	200 (260)	40 (52)	40 (52)	
Ethyl alcohol (46.07)	2,000 (3,800)	2,000 (3,800)	50 (95)	50 (95)	b
N-Butyl alcohol (74.12)	—	200 (600)	40 (120)	40 (120)	
Isobutyl alcohol (74.12)	—	200 (600)	40 (120)	40 (120)	
Sec butyl alcohol (74.12)	—	200 (600)	40 (120)	40 (120)	
Tertbutyl alcohol (74.12)	—	200 (600)	40 (120)	40 (120)	
n-Propyl alcohol (60.11)	—	200 (500)	40 (120)	40 (120)	
Isopropyl alcohol (60.11)	400 (1,000)	200 (500)	40 (100)	40 (100)	
ESTERS					
Methyl acetate (74.0)	—	200 (600)	40 (120)	40 (120)	
Ethyl acetate (88.10)	—	300 (1,080)	50 (180)	50 (180)	
Butyl acetate (116.16)	—	200 (940)	40 (188)	40 (188)	
Propyl acetate (102.1)	—	200 (840)	40 (168)	40 (168)	

TABLE 4 (Continued)

Compound (Molecular Weight)	ppm (mg/m³)				Notes
	10 Min, Special Area	60 Min	90 Days	6 Mon	
KETONES					
Acetone (58.08)	—	1,000 (2,400)	300 (720)	300 (720)	b
Methylethylketone (72.1)	—	100 (290)	20 (58)	20 (58)	
Methyl isobutylketone	—	100	20	20	
Methylisopropylketone (86.77)	—	100* (350)	20* (70)	20* (70)	c
ALDEHYDES					
Acetaldehyde (44.05)	—	50 (90)	10 (18)	10 (18)	d
Acrolein (56.06)	—	0.2 (0.5)	0.1 (0.2)	0.1 (0.2)	
Formaldehyde (30.03)	—	1.0 (1.0)	0.1 (0.1)	0.1 (0.1)	
ALICYCLICS					
Cyclohexane (82.14)	—	300 (1,020)	60 (204)	60 (204)	
Cyclopentane (70.13)	—	300 (870)	60 (204)	60 (204)	
Methyl cyclohexane (98.14)	—	500 (2,000)	15* (60)	* (60)	c
Methyl cyclopentane (84.1)	—	300 (1,029)	15* (51)	15* (51)	c

TABLE 4 (Continued)

Compound (Molecular Weight)	ppm (mg/m³)				Notes
	10 Min, Special Area	60 Min	90 Days	6 Mon	
HALOGENATED ALIPATHICS					
Chloroform (119.39)	—	100 (490)	5 (24.5)	5 (24.5)	
1,2-dichloroethane (98.97)	—	200 (800)	10 (40)	10 (40)	
Dichloromethane (84.94)	—	100 (340)	25 (87.5)	25 (87.5)	
Methyl chloroform (133.4)	—	300 (1,620)	50 (270)	50 (270)	
Tetrachloroethylene (165.85)	—	100 (680)	5 (34)	5 (34)	
R-11. Trichloroflouro-methane (140.5)	—	5,000 (28,500)	100 (570)	100 (570)	
R-12. Dichlorodiflouro-methane (124.0)	—	5,000 (25,500)	100 (510)	100 (510)	
R-113. Trichlorotriflouro-ethane (192.5)	—	500 (3,950)	50 (395)	50 (395)	
AROMATICS					
Benzene (78.11)	—	100 (320)	1.0 (3)	1.0 (3)	
Ethyl benzene (106.16)	—	200 (860)	20 (86)	20 (86)	e
Styrene (104.1)	—	50 (215)	10* (43)	10* (43)	c
Toluene (92.1)	—	200 (760)	20 (76)	20 (76)	

TABLE 4 (Continued)

Compound (Molecular Weight)	ppm (mg/m^3)				Notes
	10 Min, Special Area	60 Min	90 Days	6 Mon	
AROMATICS (continued)					
1,3,5-Trimethylbenzene (120.12)	—	25 (123)	3* (15)	3* (15)	
Xylene (o,m,p) (106.12)	—	100 (430)	20 (86)	20 (86)	
HALOGENATED AROMATICS					
Dichlorobenzene, mixed o- and p-	—	50 (300)	5 (30)	5 (30)	
HETEROCYCLICS					
1,4-Dioxane (88.0)	—	100 (360)	5 (18)	5 (18)	f
Furan (68.07)	—	2 (5)	0.04 (0.1)	0.04 (0.1)	
Indole (117.15)	1.0 (4.8)	1.0 (4.8)	0.1 (0.5)	0.1 (0.5)	e
Skatole (131.1)	1.0 (5)	1.0 (5)	0.1 (0.5)	0.1 (0.5)	e
INORGANICS					
Ammonia (17.03)	100 (70)	100 (70)	25 (17.5)	25 (17.5)	
Carbon dioxide (44.01)	40,000 (72,000)	30,000 (54,000)	10,000 (18,000)	10,000 (18,000)	
Carbon monoxide (28.01)	—	125 (144)	15 (17)	15 (17)	g,h
Hydrogen chloride, gas (36.46)	—	5.0 (7.5)	1.0 (1.5)	1.0 (1.5)	

TABLE 4 (Continued)

Compound (Molecular Weight)	ppm (mg/m³)				
	10 Min, Special Area	60 Min	90 Days	6 Mon	Notes
INORGANICS (continued)					
Hydrogen fluoride, gas (20.0)	—	5.0 (4)	0.1 (0.08)	0.1 (0.08)	
Nitrogen dioxide (46.01)	—	2.0 (4)	0.5 (1.0)	0.5 (1.0)	
Phosgene (98.92)	—	0.5 (2.0)	0.05 (0.2)	0.05 (0.2)	
Sulfur dioxide (64.1)	—	5.0 (13)	1.0 (3)	1.0 (3)	
MISCELLANEOUS					
Acetronitrile (41.05)	—	40 (68)	4.0 (6.8)	4.0 (6.8)	
Methylmercaptan (48.11)	1.0 (2)	1.0 (2)	0.1 (0.2)	0.1 (0.2)	e

[a] Adapted from NRC (1972).
[b] Not to be included in group limits.
[c] Estimated levels bear an asterisk; more inhalation data with animal models would be desirable.
[d] Based on eye irritation.
[e] Long-term limits based principally on odor.
[f] These levels for dioxane are subject to drastic revision downward (<1 ppm) if future research proves that the compound is carcinogenic in animal models at low (<100 ppm) inhalation concentrations.
[g] The 60-min limit is based on a requirement that the carboxyhemoglobin level not exceed 10%, assuming heavy-work activity (30 L/min respiration) and conformity to Coburn's equation. If the assumption of heavy-work activity in the weightless situation proves unreal, then a value of 300 ppm (330 mg/m³) is recommended.
[h] The mg/m³ limits are also specified for the 70% O_2, 30% N_2 atmosphere at 5 psia.

3 Sources of Space-Station Contaminants

Many sources of atmospheric contaminants will be present in the space station. Most sources will release only small amounts of material into the air, but contaminants may build up during operation of a closed vessel. Recognition of the many sources of atmospheric contamination has helped in eliminating major sources of contaminants and in developing methods to control and decrease contamination from any source. Table 5 lists possible space-station contaminants (Leban and Wagner, 1989). Major sources of contaminants include off-gassing of cabin materials, components, and equipment and metabolic waste products of crew members. All nonmetallic materials used in the interior of the orbiter crew compartment are known to off-gas contaminant compounds. These include cabin construction materials, electrical insulation, paints, lubricants, adhesives, and degradation of nonmetallic console and equipment structures. The heat produced by equipment operation increases off-gassing. Minor sources of contaminants in the spacecraft include internal decomposition of hydraulic fluids, electrical equipment, plastics, oil, leakage from environmental or flight control systems, volatile food components, volatile components of personal hygiene articles, and reaction products from the environmental control and life-support system (ECLSS).

Overheating of electrical components and fire can cause some structural materials (such as plastics) to emit toxic gases, vapors, and particulate matter. Pyrolysis of plastics generates a variety of contaminants, depending upon the composition of the polymer, including hydrogen chloride, carbon monoxide, hydrocyanic acid, formaldehyde, and vinyl chloride.

All structural components and other materials to be carried on board are tested to identify and quantify (under test conditions) off-gassing compounds (NASA, 1991). Off-gassing rates for each compound presented in Table 5 are based on these test results and on the quantities of each material projected to be present in the space station.

TABLE 5 Preliminary Trace-Contaminant Load Model (Eight Crew Members)[a]

Common Name	IUPAC/Accepted Name	Mol Wt, g/mol	7-day SMACs, mg/m³	Metabolic Gen Rate, mg/day[b]	Station Gen Rate, mg/day[c]	Control Method
ALCOHOLS						
Methyl alcohol	Methanol	32.04	52.40	1.42	707.000	F,M
Ethyl alcohol	Ethanol	46.07	94.00	4.00	5,216.000	F,M
Allyl alcohol	2-Propen-1-ol	58.08	1.00	0.00	3.200	F,M
Isopropyl alcohol	2-Propanol	60.11	98.30	0.00	2,022.000	F,M
n-Propyl alcohol	1-Propanol	60.11	98.30	0.00	25.300	F,M
Ethylene glycol	1,2-Ethanediol	62.07	127.00	0.00	9.500	F,M
sec-Butyl alcohol	2-Butanol	74.12	121.00	0.00	0.700	F
Isobutyl alcohol	2-Methyl-1-propanol	74.12	121.00	1.20	728.400	F
tert-Butyl alcohol	2-Methyl-2-propanol	74.12	121.00	0.00	15.800	F,M
n-Butyl alcohol	1-Butanol	74.12	121.00	1.33	6,922.000	F,M
Propylene glycol	1,2-Propanediol	76.11	0.10	0.00	0.500	F,M
n-Amyl alcohol	1-Pentanol	88.15	126.00	0.00	134.000	F
Isoamyl alcohol	3-Methyl-1-butanol	88.15	126.00	0.00	18.000	F
Carbolic acid	Phenol	94.11	7.70	0.00	7.900	F,M
Hexahydrophenol	Cyclohexanol	100.16	123.00	0.00	1,288.000	F,M
2-Ethylbutyl alcohol	2-Ethylbutanol	102.18	0.10	0.00	0.200	F
2-Hexanol	2-Hexanol	102.18	167.00	0.00	1.200	F
1,3-Dichloro-2-propanol	1,3-Dichloro-2-propanol	128.99	0.10	0.00	0.010	F
2-Ethylhexyl alcohol	2-Ethylhexanol	130.23	186.40	0.00	4.500	F
Nonyl alcohol	Nonanol	144.26	236.00	0.00	6.500	F
n-Decyl alcohol	1-Decanol	158.29	259.00	0.00	9.500	F
ALDEHYDES						
Formaldehyde	Methanal	30.03	0.12	0.00	0.020	CI

Acetaldehyde	Ethanal	44.05	54.00	0.08	48.180	F
Acrolein	Propenal	56.07	0.11	0.00	0.060	F
Propionaldehyde	Propanal	58.08	95.00	0.00	87.000	F
Methacrolein	2-Methylpropenal	70.09	0.10	0.00	0.200	F
n-Butylaldehyde	Butanal	72.12	118.00	0.00	1,470.000	F
Valeraldehyde	Pentanal	86.14	106.00	0.00	22.660	F
Sorbaldehyde	2,4-Hexadien-1-al	96.14	4.70	0.83	1.500	F
Caproaldehyde	Hexanal	100.16	4.70	0.00	43.000	F
Benzenecarbonal	Benzaldehyde	106.13	173.00	0.00	43.000	F
Enanthaldehyde	Heptanal	114.19	0.10	0.00	11.000	F
p-Tolualdehyde	4-Methylbenzaldehyde	120.16	0.10	0.00	4.000	F
Caprylaldehyde	Octanal	128.22	210.00	0.00	3.200	F

AROMATIC HYDROCARBONS

Benzene	Benzene	78.12	0.32	0.00	27.000	F
Toluene	Methylbenzene	92.15	75.30	0.00	1,351.000	F
Styrene	Ethenylbenzene	104.16	42.60	0.00	9.500	F
p-Xylene	1,4-Dimethylbenzene	106.17	86.80	0.00	780.000	F
Ethylbenzene	Ethylbenzene	106.17	86.80	0.00	182.000	F
o-Xylene	1,2-Dimethylbenzene	106.17	86.80	0.00	106.00	F
m-Xylene	1,3-Dimethylbenzene	106.17	86.80	0.00	3,539.000	F
Indonaphthene	Indene	116.16	9.50	0.00	118.000	F
α-Methylstyrene	α-Methylstyrene	118.18	145.00	0.00	1.200	F
1-Ethyl-2-methylbenzene	1-Ethyl-2-methylbenzene	120.20	25.00	0.00	5.000	F
Pseudocumene	1,2,4-Trimethylbenzene	120.20	15.00	0.00	16.000	F
Propylbenzene	Propylbenzene	120.20	49.10	0.00	269.00	F
Cumene	Isopropylbenzene	120.20	73.70	0.00	11.000	F
Mesitylene	1,3,5-Trimethylbenzene	120.20	15.00	0.00	2.000	F
n-Butylbenzene	n-Butylbenzene	134.22	55.00	0.00	2.400	F
m-Cymene	1-Methyl-3-propylbenzene	134.22	11.00	0.00	2.700	F
p-Cymene	1-Methyl-4-propylbenzene	134.22	0.10	0.00	0.500	F

TABLE 5 (Continued)

Common Name	IUPAC/Accepted Name	Mol Wt, g/mol	7-day SMACs, mg/m^3	Metabolic Gen Rate, mg/day[b]	Station Gen Rate, mg/day[c]	Control Method
ESTERS						
Methyl formate	Methyl formate	60.05	12.30	0.00	0.060	F
Ethyl formate	Ethyl formate	74.08	90.90	0.00	0.800	F
Methyl acetate	Methyl acetate	74.08	121.00	0.00	11.000	F
Ethyl acetate	Ethyl acetate	88.12	180.00	0.00	371.000	F
Ethyl cellosolve	2-Ethoxyethanol	90.12	73.70	0.00	1,035.000	F,M
Methyl methacrylate	Methyl-2-methyl propenoate	100.13	102.00	0.00	24.000	F
Allyl acetate	Allyl acetate	100.13	51.20	0.00	5.000	F
n-Butyl formate	n-Butyl formate	102.13	83.50	0.00	0.050	F
Isopropyl acetate	Isopropyl acetate	102.13	209.00	0.00	3.200	F,M
Propyl acetate	Propyl acetate	102.13	167.00	0.00	585.000	F
Ethyl methacrylate	Ethyl-2-methyl propenoate	114.15	116.70	0.00	36.000	F
Butyl acetate	Butyl acetate	116.16	190.00	0.00	948.000	F
Isobutyl acetate	Isobutyl acetate	116.16	190.00	0.00	245.000	F
Ethyl lactate	Ethyl-2-hydroxy propanoate	118.13	193.00	0.00	245.000	F,M
Methyl cellosolve acetate	2-Methoxy ethyl acetate	118.13	24.20	0.00	5.000	F
Isoamyl acetate	3-Methylbutyl acetate	130.19	159.50	0.00	3.200	F
n-Amyl acetate	n-Pentyl acetate	130.19	160.00	0.00	79.000	F
Cellosolve acetate	2-Ethoxyethyl acetate	132.16	162.00	0.00	545.000	F
Ethyl acetoxyacetate	Ethyl acetylglycolate	146.14	0.10	0.00	0.700	F
Dibutyl oxalate	Dibutyl oxalate	202.25	0.10	0.00	0.040	F
ETHERS						
Furan	1,4-Epoxy-1,3-butadiene	68.08	0.11	0.00	1.600	F
Tetrahydrofuran	1,4-Epoxybutane	72.12	118.00	0.00	95.000	F
Allyl methyl ether	3-Methoxy-1-propene	72.12	0.10	0.00	0.060	F

Ether	Diethyl ether	74.12	242.00	0.00	52.000	F
Sylvan	2-Methylfuran	82.10	0.13	0.00	1.000	F
Gluconal	2,3-Dihydropyran	84.13	0.10	0.00	0.400	F
p-Dioxane	1,4-Dioxane	88.12	1.80	0.00	63.000	F,M
Metaformaldehyde	1,3,5-Trioxane	90.08	0.10	0.00	0.020	F,M
β-Epichlorohydrin	2-Chloro-1,3-epoxypropane	92.53	1.20	0.00	5.000	F,M
Tetramethyl oxirane	Tetramethyl-1,2-epoxy-ethane	100.12	0.10	0.00	1.600	F,M
N-Ethyldiethylenimide oxide	4-Ethylmorphone	115.18	16.00	0.00	213.000	F
Butyl propyl ether	1-Propoxybutane	116.21	186.80	0.00	55.000	F
Glycol monobutyl ether	2-Butoxyethanol	118.18	24.20	0.00	0.005	F,M

HALOCARBONS - CHLOROCARBONS

Methyl chloride	Chloromethane	50.49	41.30	0.00	0.300	C,F
Vinyl chloride	Chloroethene	62.50	0.26	0.00	1.600	F
Ethyl chloride	Chloroethane	64.52	263.70	0.00	545.000	F
Allyl chloride	3-Chloropropene	76.53	0.63	0.00	34.000	F
Methylene chloride	Dichloromethane	84.93	86.80	0.00	1,746.000	C,F
n-Butyl chloride	1-Chlorobutane	92.57	151.00	0.00	3.000	F
Dichloroethene	1,1-Dichloroethene	96.95	7.90	0.00	0.020	F
Ethylene dichloride	1,2-Dichloroethane	98.97	40.50	0.00	20.000	F
1,3-Dichloropropylene	1,3-Dichloropropene	106.97	42.20	0.00	47.000	F
Chlorobenzene	Chlorobenzene	112.56	46.00	0.00	1,240.000	F
Propylene chloride	1,2-Dichloropropane	112.99	42.20	0.00	4.000	F
Chloroform	Trichloromethane	119.38	4.90	0.00	9.500	F
Isobutylene chloride	1,2-Dichloro-2-methyl-propane	127.01	0.10	0.00	0.800	F
Trichloroethylene	Trichloroethene	131.39	0.54	0.00	40.000	F
Vinyl trichloride	1,1,2-Trichloroethane	133.41	5.50	0.00	2.400	F
Methyl chloroform	1,1,1-Trichloroethane	133.41	164.00	0.00	229.000	F
o-Dichlorobenzene	1,2-Dichlorobenzene	147.01	30.00	0.00	11.000	F
Chloromethylheptane	3-Chloromethylheptane	148.68	0.10	0.00	0.300	F

TABLE 5 (Continued)

Common Name	IUPAC/ Accepted Name	Mol Wt, g/mol	7-day SMACs, mg/m^3	Metabolic Gen Rate, mg/day^b	Station Gen Rate, mg/day^c	Control Method
HALOCARBONS - CHLOROCARBONS (continued)						
Carbon tetrachloride	Tetrachloromethane	153.82	13.00	0.00	1.600	F
Tetrachloroethylene	Tetrachloroethene	165.83	34.00	0.00	553.00	F
HALOCARBONS - FLUOROCARBONS						
Freon 22	Chlorodifluoromethane	86.47	353.60	0.00	467.000	C,F
Freon 21	Dichlorofluoromethane	102.90	21.00	0.00	5.500	F
Chlorotrifluoroethane	1-Chloro-1,2,2-trifluoro-ethane	118.50	484.50	0.00	2.400	F
Freon 12	Dichlorodifluoromethane	120.91	494.40	0.00	14.000	F
Dichlorodifluoroethene	1,2-Dichloro-1,2-difluoro-ethene	132.93	136.00	0.00	0.800	F
Freon 124	Chlorotetrafluoroethane	136.48	555.00	0.00	750.00	F
Freon 11	Trichlorofluoromethane	137.40	561.80	0.00	174.00	F
Halon 1301	Bromotrifluoromethane	148.90	608.80	0.00	474.00	F
Freon 114	1,1-Dichloro-1,2,2,2-tetra-fluoroethane	170.92	702.90	0.00	0.010	
Freon 113	1,1,2-Trichloro-1,2,2-tri-fluoroethane	187.40	383.00	0.00	22,981.000	F
Freon 112	1,1,2,2-Tetrachloro-1,2-di-fluoroethane	204.00	834.20	0.00	103.000	F
HYDROCARBONS						
Methane	Methane	16.04	1,771.00	600.00	1,620.00	C
Acetylene	Ethyne	26.04	532.40	0.00	26.000	C
Ethylene	Ethene	28.05	344.10	0.00	0.400	C
Ethane	Ethane	30.07	1,230.00	0.00	166.000	C,F

Name	Synonym	MW				
Allene	Propadiene	40.07	81.90	0.00	180.000	C,F
Methylacetylene	Propyne	40.07	409.50	0.00	8.700	F
Propylene	Propene	42.08	860.30	0.00	0.500	F
Propane	Propane	44.09	901.40	0.00	0.500	F
Vinylethylene	1,3-Butadiene	54.09	221.20	0.00	4.700	F
Ethylethylene	1-Butene	56.10	458.00	0.00	40.000	F
Isobutane	2-Methylpropane	58.12	237.60	0.00	95.000	F
Butane	Butane	58.12	237.60	0.00	1.600	F
Isoprene	2-Methyl-1,3-butadiene	68.12	557.00	0.00	148.000	F
Cyclopentene	Cyclopentene	68.13	167.00	0.00	130.000	F
Propylethylene	1-Pentene	70.13	186.00	0.00	35.000	F
Pentane	Pentane	72.15	590.00	0.00	134.000	F
Isopentane	2-Methylbutane	72.15	295.00	0.00	3.200	F
Cyclohexene	1,2,3,4-Tetrahydrobenzene	82.14	86.00	0.00	35.000	F
Acetyl cyclopropane	Acetyl cyclopropane	84.13	0.10	0.00	0.080	F
Methylpentamethylene	Methylcyclopentane	84.16	51.60	0.00	51.000	F
2-Hexene	2-Hexene	84.16	172.00	0.00	0.800	F
Hexamethylene	Cyclohexane	84.16	206.00	0.00	624.000	F
Neohexane	2,2-Dimethylbutane	86.17	88.10	0.00	3.200	F
Hexane	Hexane	86.18	176.00	0.00	81.000	F
Diethylmethylmethane	3-Methylpentane	86.18	1,762.00	0.00	2.800	F
Methylcyclohexene	4-Methylcyclohexene	96.17	393.20	0.00	253.000	F
Hexahydrotoluene	Methylcyclohexane	98.18	60.20	0.00	69.000	F
1-Heptylene	1-Heptene	98.18	201.00	0.00	113.000	F
Triethylmethane	3-Ethylpentane	100.21	201.00	0.00	0.100	F
2,2-Dimethylpentane	2,2-Dimethylpentane	100.21	408.60	0.00	86.000	F
2,4-Dimethylpentane	2,4-Dimethylpentane	100.21	201.00	0.00	0.400	F
Heptane	Heptane	100.21	201.00	0.00	79.000	F
6-Methyl-1-heptene	6-Methyl-1-heptene	112.22	229.00	0.00	0.020	F
Hexamethylene	trans-1,2-Dimethylcyclohexane	112.22	115.00	0.00	95.000	F

TABLE 5 (Continued)

Common Name	IUPAC/ Accepted Name	Mol Wt, g/mol	7-day SMACs, mg/m³	Metabolic Gen Rate, mg/day[b]	Station Gen Rate, mg/day[c]	Control Method
HYDROCARBONS (continued)						
Dimethylcyclohexane	1,1-Dimethylcyclohexanehe	112.22	115.00	0.00	24.000	F
2-Octene	2-Octene	112.22	229.00	0.00	22.000	F
2,2,3-Trimethylpentane	2,2,3-Trimethylpentane	114.23	229.00	0.00	0.020	F
3-Ethylhexane	3-Ethylhexane	114.23	229.00	0.00	95.000	F
3,3-Dimethylhexane	3,3-Dimethylhexane	114.23	229.00	0.00	24.000	F
Octane	Octane	114.23	350.00	0.00	66.000	F
Nonane	Nonane	128.26	315.00	0.00	1.700	F
4-Ethylheptane	4-Ethylheptane	128.26	129.00	0.00	0.060	F
Citrene	4-Isopropenyl-1-methyl-cyclohexene	136.23	557.00	0.00	6.000	F
Decane	Decane	142.28	223.00	0.00	0.080	F
Methylethylheptane	2-Methyl-3-ethylheptane	142.28	116.00	0.00	0.080	F
Undecane	Undecane	156.31	319.00	0.00	14.000	F
Dodecane	Dodecane	170.34	278.00	0.00	5.500	F
KETONES						
Acetone	2-Propanone	58.08	712.50	0.13	4,212.400	F,M
Methyl vinyl ketone	3-Buten-2-one	70.00	0.10	0.00	0.300	F
Methyl ethyl ketone	2-Butanone	72.11	59.00	0.00	3,760.000	F
Adipic ketone	Cyclopentanone	84.11	29.20	0.00	845.000	F
Methyl propenyl ketone	3-Penten-2-one	84.12	0.10	0.00	0.050	F
Cyclopropyl methyl ketone	Acetyl cyclopropane	84.13	0.10	0.00	0.081	F,M
Methyl propyl ketone	2-Pentanone	86.13	70.40	0.00	4.700	F
Methyl isopropyl ketone	3-Methyl-2-butanone	86.13	70.40	0.00	4.700	F

Pimelic ketone	Cyclohexanone	98.14	60.20	0.00	292.000	F
Methyl isobutenyl ketone	4-Methyl-3-penten-2-one	98.15	40.10	0.00	47.000	F
Methyl isobutyl ketone	4-Methyl-2-pentanone	100.16	82.00	0.00	1,335.000	F
Pinacolone	3,3-Dimethyl-2-butanone	100.16	81.90	0.00	6.300	F
Methyl isoamyl ketone	5-Methyl-2-hexanone	114.18	23.50	0.00	0.600	F
Diisopropyl ketone	2,4-Dimethyl-3-pentanone	114.18	23.50	0.00	2.400	F
Methyl *n*-amyl ketone	2-Heptanone	114.18	23.50	0.00	0.400	F,M
Phenyl methyl ketone	Acetophenone	120.14	245.00	0.00	1.600	F
Methyl hexyl ketone	2-Octanone	128.21	105.00	0.00	0.300	F
Methylheptanone	5-Methyl-3-heptanone	128.21	0.10	0.00	1.600	F
Diisobutyl ketone	2,6-Dimethyl-4-heptanone	142.20	58.10	0.00	711.000	F

MERCAPTANS AND SULFIDES

Hydrogen sulfide		34.08	0.08	0.00	0.700	C
Methyl mercaptan	Methanethiol	48.11	0.20	0.83	0.830	F
Ethylene sulfide		60.07	0.10	0.00	0.060	F
Carbon oxisulfide	Carbonyl sulfide	60.07	12.00	0.00	5.400	F,C
Ethyl mercaptan	Ethanethiol	62.13	0.25	0.83	0.830	F
Methyl sulfide	Dimethyl sulfide	62.14	2.50	0.00	0.300	F
Carbon disulfide		76.14	16.00	0.00	44.000	F
n-Propyl mercaptan	1-Propanethiol	76.17	0.10	0.83	0.830	F
Tetrahydropyran	Pentamethylene sulfide	102.20	0.10	0.00	0.080	F

NITROGEN OXIDES

Nitric oxide		30.01	6.10	0.00	0.044	PU
Nitrogen dioxide		46.01	0.94	0.00	0.020	F,L
Nitrogen tetroxide		92.01	1.90	0.00	48.000	F,M

31

TABLE 5 (Continued)

Common Name	IUPAC/ Accepted Name	Mol Wt, g/mol	7-day SMACs, mg/m³	Metabolic Gen Rate, mg/day[b]	Station Gen Rate, mg/day[c]	Control Method
ORGANIC ACIDS						
Acetic acid	Ethanoic acid	60.05	7.40	0.00	0.020	F,M,L
Pyruvic acid	2-Oxopropanoic acid	88.06	3.59	208.30	208.300	F
Valeric acid	n-Pentanoic acid	102.13	103.98	0.83	0.830	F
Ethylhexanoic acid	2-Ethylhexanoic acid	144.21	0.10	0.00	0.600	F
Caprylic acid	Octanoic acid	144.22	146.55	9.17	9.170	F
ORGANIC NITROGENS						
Acetonitrile	Methyl cyanide	41.05	6.70	0.00	83.000	F,M
Monomethyl hydrazine	Methyl hydrazine	46.07	0.08	0.00	2.400	F
Nitromethane	Nitromethane	61.04	0.10	0.00	8.000	F
N,N-Dimethylformamide	N,N-Dimethylformamide	73.10	6.00	0.00	0.600	F,M
Nitroethane	Nitroethane	75.07	0.10	0.00	0.020	F
Indole	1-Benzopyrrole	117.15	0.48	25.00	100.000	F
Skatole	3-Methylindole	131.18	0.54	25.00	25.000	F
MISCELLANEOUS - INORGANIC						
Hydrogen	Hydrogen	2.02	247.30	50.00	208.000	C
Ammonia	Ammonia	17.00	17.40	250.00	3,806.000	PA
Carbon monoxide	Carbon monoxide	28.01	28.60	33.30	1,843.000	CO,C
Diamine	Hydrazine	32.05	0.05	0.00	1.680	F,M
Mercury	Mercury	200.59	0.006	0.00	1.200	F
MISCELLANEOUS - SILANES AND SILOXANES						
Siloxane dimer	Siloxane dimer	78.10	52.40	0.00	32.000	F

Trimethylsilanol	90.21	1.80	12.000	F
Siloxane trimer	124.30	83.40	24.000	F
Hexamethyldisiloxane	162.48	96.60	0.080	F
Siloxane tetramer	170.40	114.00	237.000	F
Diphenyl silane	184.32	0.10	0.020	F
Hexamethylcyclotrisiloxane	222.40	227.00	47.000	F
Octamethyltrisiloxane	236.54	114.00	237.000	F
Octamethylcyclotetrasiloxane	296.62	151.70	71.000	F
Decamethylcyclopentasiloxane	370.64	150.70	316.000	F
Decamethylcyclohexasiloxane	444.71	150.70	403.00	F
Tetradecamethylcycloheptasiloxan	519.09	150.70	555.000	F
Hexadecamethylcyclooctasiloxane	593.24	150.70	126.000	F

[a]From Leban and Wagner (1989).
[b]Metabolic generation rate for one person.
[c]Predicted Space Station Freedom (SSF) generation rate for four modules and eight crew members. Calculated by taking space-lab data. Samples were taken before the space-lab mission (fully outfitted). SSF was calculated by:

space-lab generation rate × ratio of SSF wt. Wt ratio = 7.9.
                                    space-lab wt

Contaminant removal methods:

F = Fixed activated charcoal bed.
M = Sorption by moisture (condensing heat exchanger).
PA = Sorption by phosphoric acid (impregnated charcoal).
L = Sorption by $LiOH/Li_2CO_3$ (pre- and post-sorber).

CO = Low temperature catalytic oxidizer.
C = High temperature catalytic oxidizer.
CI = CI-type charcoal.
PU = Purafil.

Escape of liquid or gaseous chemicals from water reclamation, cooling, and propellant systems; combustion; thermal decomposition; and vaporization could lead to a major contamination event. For these reasons, only ammonia cooling loops and hydrazine propellants will be permitted to be located outside the pressurized modules.

Complete and multiple containments of processing and experimental apparatus are required by NASA if hazardous chemicals are present. Nevertheless, the possibility of accidental release of chemical, radioactive, or biological agents cannot be completely ruled out. Such releases could produce serious contamination and high concentrations in space-station air. Some of the commercial-type facilities expected to be used could accidentally release a significant amount of chemicals.

In the space station, cleaning and degreasing agents, adhesives, disinfectants, and lubricants will be used in cleaning and maintenance procedures. Another potential source of contamination is repair of metallic or plastic objects involving soldering, welding, or drilling. These repair processes often involve localized heating and may release toxic gases, vapors, or particles.

Chemical and physical processes in cabin air involving sunlight (high-energy solar radiation) and some airborne contaminants may result in the formation of highly toxic chemicals such as ozone, nitrogen dioxide, and other photochemical products (peroxyacetyl nitrate, hydrogen peroxide, or free radicals). The formation of such products is a complex, nonlinear function involving many factors including the intensity and spectral distribution of sunlight, the concentration of the many precursors (aldehydes or carbon monoxide), the ratio between organic compounds and the oxides of nitrogen, and the reactivity of organic precursors. One resulting product, ozone, is highly toxic at low concentrations, causing a number of pulmonary and extrapulmonary effects. Due to the highly toxic nature of this gas, the potential hazard associated with such exposure in the spacecraft environment is a cause for significant concern.

The principal sources of contaminants from metabolic waste products of humans are urine, feces, flatus, and expired air. Animal urine and dander are potential sources of contaminants; however, NASA believes they are not a problem.

Urine will be a contaminant in the space station only if the design is found to be ineffective in handling urine, leading to leakage or spillage. Urinary contaminants would be only a nuisance; in the absence of an infection of the urinary tract, urine is sterile.

With feces, again the problem is proper mechanical and personal handling. Experience in spaceflights has indicated that bowel regularity (one or two stools per day) facilitates management. A method of vacuum cleaning of air contaminants has been devised in addition to surface cleaning.

Flatus is an appreciable source not only of periodic discomfort due to odor but also of toxic gases, which must be removed from the cabin air. As studied by Calloway (1971), flatus of men fed a low-residue formula diet was 104 mL/12 hr (mean) in contrast to 286 mL/12 hr from men fed a diet of processed foods similar to those given to astronauts in the Gemini flight series. When gas-producing foods are eaten, flatus passage may increase to 60–120 mL/hr (Calloway, 1968). Small volumes of flatus tend to have a composition like (ingested) air, whereas large volumes contain more carbon dioxide and flammable gases produced by bacteria, principally hydrogen and methane. These gases, in addition to being expelled, are absorbed into the bloodstream and then appear in expired air. Calloway and Murphy (1969), by measuring both expired air and flatus, found potential total daily volumes of 730 mL of hydrogen and 380 mL of methane in men on a spaceflight-type diet. Although the investigators raised the possibility that these amounts of hydrogen and methane "could constitute a fire hazard in a closed chamber," it seems unlikely that, in the sizeable space planned for the space station and with functional environmental controls, concentrations of these compounds would be high enough for risk of explosion. The malodorous compounds in flatus are indole, skatole, mercaptans, ammonia, and hydrogen sulfide. The amounts of these compounds and others in flatus depend on two principal factors, the numbers and kinds of enteric organisms present and the substrate for these bacteria provided by diet. The importance of minimizing the number and amount of gas-producing foods for space diets is evident. In any event, the atmosphere of the space station will be continually cleansed.

The principal expired respiratory gas is carbon dioxide at 15–20 L/hr, though possibly as much as four or five times that in extravehicular exertion (Calloway, 1971). Small amounts of carbon monoxide, hydrogen, methane, and ammonia are also expired.

Finally, skin cells are in a constant state of turnover, and there is a constant daily flaking of desiccated dead-cell remnants. This dusty material, though relatively minor in amount, would be a contaminant if a regular washing and wiping routine were not followed by astronauts in the space station. Studies of normal subjects (Calloway, 1971) have shown that several nitrogen-containing and organic compounds and minerals are lost from the skin with insensible perspiration and still more with active sweating. Measurements of actual losses from the skin in space are limited to those taken on the Gemini VII spaceflight (Lutwak et al., 1969). Mineral losses were found to be minor during this 14-day flight. Sweating was minimal, as would be expected because most astronaut activities, although they require good neuromuscular coordination, need little physical exertion except during extravehicular operations. Since mineral and other losses from the skin increase into the significant range only with active sweating, the possibility of contamination from metabolic losses from the skin will be reduced by minimizing the likelihood of sweating, presumably by suitable temperature control and by moderation in physical activity.

There is a possibility of potential contamination from dust mites, which produce proteins that are potent allergens. However, according to NASA, dust mites have not proved to be a problem.

The water reclamation system in the space station may be a potential source of toxic chemicals in the cabin air. The system may contribute several types of contaminants to space-station air.

1. Chemicals originally collected in the cabin condensate and then revolatilized.
2. Components of hygiene wastewater.
3. Chemicals actually produced in water treatment.

Reclaimed water after treatment is the major source of drinking water. Any chemicals present in the air will obviously be condensed with water in proportion to their partial pressure and water solubility. For most nonpolar volatile chemicals, cycling through the water system is likely to contribute little to the hazards first encountered in the air. On the other hand, chemicals that are both polar and volatile (e.g., alcohols) may concentrate and even accumulate in the water system. Under such circumstances, the major portion of the exposure to the chemical may come via water.

Hygiene wastewater will contain soaps, detergents, and other chemicals present in personal hygiene products that might be expected from showering and laundering clothing. Most of these chemicals are relatively nonvolatile and unlikely to contaminate the air. These chemicals are removed fairly efficiently by water treatment processes.

The water reclamation system has the potential to introduce novel compounds into the air. The use of oxidants to disinfect and chemically degrade water contaminants is the major means by which such chemicals are produced. For example, the use of persulfate and sulfuric acid to treat urine results in the generation of cyanogen chloride. If this reaction product were to elude subsequent treatment processes, it could introduce a sufficient amount into the confined space of the shower to produce acute respiratory irritation. Repeated exposure to such irritants over a 90- to 180-day period could have long-term effects. Similar but less well-defined problems might result from the use of iodine (or other reactive chemical) as the residual disinfectant in the water system.

Crews from several space-shuttle flights have reported eye and respiratory tract irritation associated with the presence of airborne particles and floating debris in the shuttle cabin. The debris included paint chips, metal shavings, food particles, and fibrous materials, including fibers from clothing, paper wipes, and fiberglass.

A panel assembled by NASA on Airborne Particulate Matter in Spacecraft (NASA, 1988) recommended that particle concentration in the cabin should be "as low as reasonably achievable" (ALARA). The panel noted that simple technology could be used to discern the particle concentration in the atmosphere because the particles are likely to have many of the characteristics of nuisance dust. The panel questioned whether the reported symptoms of eye and respiratory tract irritation were due to those particles. The symptoms also could result from exposure to gases such as nitrogen dioxide, ozone, or formaldehyde present in the air.

The recommendations of ALARA particle levels should allow, however, for design of appropriate particle-control technology. The panel recommended the following numerical limits:

1. For flights of 1 week or less: 1 mg/m^3 limit for particles <10 μm in AD (aerodynamic diameter) + 1 mg/m^3 for particles 10-100 μm in AD.
2. For flights of >1 week and up to 6 months: 0.2 mg/m^3 for particles 10-100 μm in AD.

No specific limit for particles >100 μm in AD was recommended because adequate particle cleanup to meet the above values should result in acceptable levels of larger particles.

In recommending these limits, the panel considered acute and chronic irritation of the respiratory tract and eyes to be the primary concerns. In selecting the 1-mg/m^3 limit for short flights, the panel considered that if the irritation reported was due to particles and not gases, this limit should protect from irritation of the respiratory tract. This exposure limit was based on data from exposure of healthy humans to submicrometer-sized aerosols of sulfuric acid at concentrations as high as 1 mg/m^3 with no signs of respiratory tract or eye irritation. The limit for flights of longer duration was lowered by a factor of five to allow for the uncertainties about the toxicity of the particles. The 0.2-mg/m^3 limit is the same as that set for U.S. submarines where conditions are somewhat similar to spacecraft.

Obviously, exposure to chemicals from water and air are not neatly separable problems. Therefore, it is important that assessments of risks from certain concentrations of chemicals in air include consideration of the potential risks associated with possible buildup of the chemicals in the potable water system. Attention should be paid to defining the toxic effects and exposure scenarios for chemicals that are confined in the water system only.

Sources of contaminants have been categorized into three subgroups according to predictability of release. Category 1 sources are those that release contaminants continuously or frequently or are associated with a specific routine activity. A principal characteristic of sources in this category is that their contaminant-generation rates can be predicted with a high degree of accuracy. Systems usually can be designed to keep exposures to compounds emitted from these sources at or below the SMAC values. The emission rates given in Table 5 are based on sources of this type.

Category 2 sources involve events such as inadvertent, accidental, or emergency releases of contaminants. Events leading to such releases include leaks, spills, failure of storage vessels, and overheating of components. Contaminants that may be released from category 2 sources include all gases and liquids normally kept on the space station and the thermal and chemical breakdown products of solid materials (for example, electrical insulation). Unlike category 1 sources, potential release rates of category 2 sources cover many orders of magnitude, even for a single source. Thus, planning for control of such releases involves examining various failure scenarios. Reliability analysis to establish the likelihood of failure events is an important step in the risk assessment of contaminants from category 2.

Category 3 sources are those involving accidental release from experiments performed on the space station. Some possible contaminants may be identified by an examination of the manifests of the materials for the experiments. Other contaminants may result from chemical reactions within an experimental module; recognition of such contaminants requires a detailed understanding of the module and the experimental work or production to be carried out. Because every effort will be made to eliminate releases from experimental modules through the use of triple-containment systems, almost all releases from category 3 sources, as from category 2, are likely to be accidental. Unlike category 2, it is not possible to foresee with any certainty the contaminants that may be released from these sources or the magnitude of release before space-station design and launch because many experiments will be performed over the space station's lifetime. Novel compounds may be released in the space station (e.g., new alloys or crystals during water recycling). Evaluation of such compounds usually will be required after identification because toxicity data on new compounds frequently are not available.

4 Environmental Control and Life-Support System

The environmental control and life-support system (ECLSS) is designed to control the temperature, humidity, and composition of space-station air; recover water; dispose of waste; and detect and suppress fires. ECLSS consists of six subsystems: temperature and humidity control (THC), atmosphere control and supply (ACS), atmosphere revitalization subsystem (ARS), fire detection and suppression (FDS), water recovery and management (WRM), and waste management (WM). The first three subsystems are directly related to the maintenance of cabin air quality.

The THC subsystem controls cabin air temperature, humidity, and recirculation rate, as well as the exchange of air between modules and equipment air cooling. This subsystem also includes food storage and control of airborne particulates and microbes. Air temperature and humidity are controlled by a condensing heat exchanger with automatically controlled bypasses in each node and module of the station. From a central location, the THC subsystem manages intermodule ventilation, air exchange between pressured elements of the station to maintain the proper total pressure and O_2 and CO_2 partial pressures. Particulate contaminants are to be controlled by drawing the cabin atmosphere through filter elements that consist of 70-μm, hydrophilic, pleated screens followed by dimple-pleated, borosilicate-fiber HEPA filters. A protective metal grate collects much larger particles before they enter the filter element.

The ACS subsystem will use stored O_2 and N_2 for contingency repressurization and compensation for normal atmospheric leakage. The ARS is intended to monitor and control the major components of air (O_2, N_2, H_2O, and CO_2) within the pressurized portions of the space station. H_2, CO, and CH_4 concentrations also will be monitored. The ARS functions consist of atmospheric monitoring, CO_2 removal, and contaminant control. Removal of trace contaminants is to be accomplished by the trace contaminant control subassembly (TCCS) that consists of a fixed activated charcoal bed, a high-temperature catalytic oxidizer, and a lithium hydroxide bed. In the baseline subsystem, CO_2 will be removed by a four-bed molecular sieve.

CONTAMINANT CONTROL AND MONITORING

The schematic in Figure 1 shows the portions of the ECLSS devoted to contaminant control for a single module. Air is to be circulated at a relatively high rate through the cabins, and large debris and small particles will be removed at each cabin's return air vent. Approximately 60% of this air stream will be recirculated back to the cabin and 40% will pass through a condensing heat exchanger to remove excess moisture. The condensate will be a source of potable water. Assuming 100% removal efficiency and a module volume of 74.23 m^3, the TCCS has an equivalent dilution capacity of 0.21 air changes per hour for contaminants removed by charcoal (at a charcoal-bed flow rate of 15.29 m^3/hr) and 0.06 air changes per hour for contaminants removed by catalytic oxidation only (at an oxidizer flow rate of 4.25 m^3/hr).

The ARS is designed to maintain exposures below the 180-day SMACs for normal rates of contaminant generation when the station is "permanently manned," and below the 30-day SMACs when the station is only "man-tended." In addition, means are being devised to reduce contaminant levels following unanticipated releases to meet the 1- and 24-hr SMACs, but these additional safety measures are not part of the baseline design.

Air monitoring is critical to the functioning of the ACS subsystem and the ARS (Humphries et al., 1990), not to mention to the astronauts' health and safety. The ARS includes the major constituent analyzer (MCA). This will be a mass spectrometer for analyzing cabin air for O_2, N_2, H_2, CO_2, H_2O, and CH_4 in all pressurized areas of the habitation and laboratory modules. Air samples will be drawn to the MCA for analysis from seven locations sequentially. Each analysis will require about 1 min. Thus, measurements will be made at each location approximately once every 8 min. A separate instrument will monitor CO by nondispersive infrared spectroscopy.

The ARS includes the atmospheric composition monitor assembly (ACMA), which will measure trace contaminants, CO, total particulates, and the major atmospheric constituents listed above. Trace

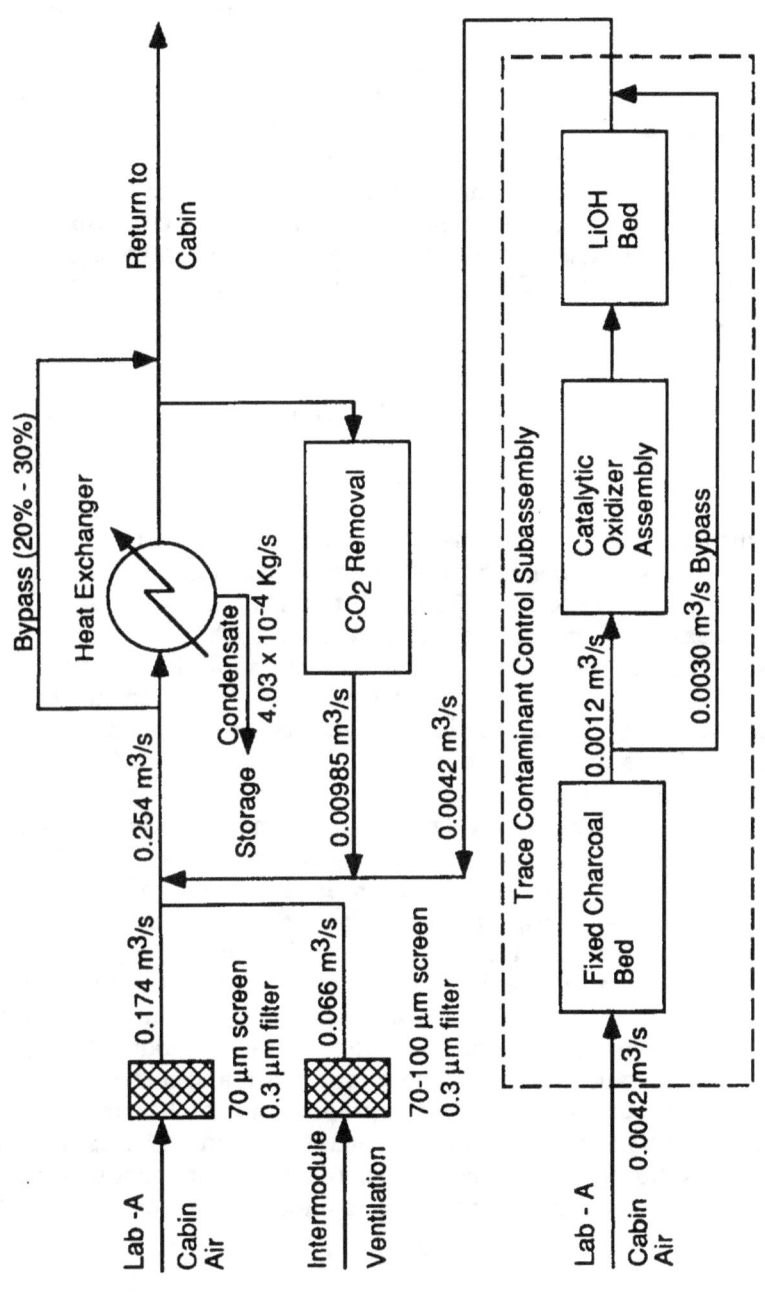

FIGURE 1 Lab-A. An air-flow diagram of the subsystems THC and ARS.

contaminants will be analyzed by gas chromatography/mass spectroscopy. As in the MCA, the ACMA will bring air samples from pressurized areas to the central ECLSS analytical assembly. Each analysis will require approximately 30 min, resulting in a monitoring frequency at each location of about once every 4 hr.

In addition to the analytical instrumentation in the ECLSS, the toxicology subsystem of the environmental health system for crew health care has extensive air-monitoring capability and also includes dedicated continuous-monitoring instruments for total hydrocarbons, hydrazine, HCl, HF, HCN, CO, and 30 targeted volatile organic compounds.

RECOMMENDATIONS

Sources

Criteria for permitting materials on the space station should be developed based on their physical, chemical, and toxicological properties and the ability of the ARS to limit their concentrations. If some experiments utilizing substances that do not meet these criteria are judged to be of such high priority that they should be performed on the space station, the experimental modules should be subjected to extensive testing, preferably in microgravity, to ensure complete containment. The subcommittee recommends development of instruments for the continuous monitoring of contaminants of concern.

The subcommittee strongly recommends that the space-station design include the capability of isolating the laboratory modules from other portions of the space station in the event of severe contamination. Good design practices for laboratories on earth usually include restricting air flow from the laboratory to adjoining nonlaboratory spaces by keeping the laboratory under negative pressure with respect to those adjoining spaces. On the space station, however, air will be flowing, by design, from the laboratory module to adjacent nodes to maintain CO_2 levels and temperature and humidity control. An additional consideration is that keeping the hatches open between modules and nodes facilitates astronaut movement, creating more efficient working conditions. Thus, the laboratory modules are not isolated from the crew living quarters under normal conditions. Because of the forced air exchange and open hatches, accidental release of an acutely toxic material in the laboratory module could contaminate air throughout the space station before measures could be taken to close off the laboratory area. Some toxicants work so rapidly that those exposed at high levels are incapable of donning their emergency breathing apparatus.

In the current design, CO_2 is to be used as a fire suppressant. Halon 1301, an alternative agent used on the space shuttle, presents little problem when air is only partially recirculated. It is not recommended for complete recirculation because the ARS will remove chlorofluorocarbons only very slowly and may generate more toxic airborne contaminants. The subcommittee is concerned also that some products generated by fires and by chemical fire extinguishers could rapidly and severely compromise performance and impair the health of the astronauts.

Rates of emission of contaminants have been estimated for category 1 sources, i.e., those that emit contaminants continuously or routinely. For sources in categories 2 and 3 (failure events and accidental releases from experimental modules), reliability analysis should be performed to estimate the likelihood of release. For example, overheating of an electrical component can lead to a high rate of emission from electrical insulation. The probability of such events should be estimated and the potential contaminants and release rates determined. This information may then be incorporated in the risk assessment. Appropriate countermeasures should be developed for such events if the baseline ARS is shown to be inadequate.

Environmental Control and Life-Support System

The design criteria for the ARS have been driven largely by the need for the control of contaminants from category 1 sources. The primary concern has been to keep contaminants that are released continuously or frequently at acceptable levels. More consideration needs to be given to control of contaminants from accidental releases. Based on the volume of Lab-A module (74.23 m^3), Figure 1 shows that about one-fifth of the module's atmosphere passes through the TCCS in 1 hr.

Assuming 100% removal efficiency, a module volume of 74.23 m^3, and the flow rates given in Figure 1, the time required to achieve various fractional reductions in the concentrations after a sudden release is shown below for contaminants removed by the charcoal bed and for contaminants removed only by the catalytic oxidizer.

Fractional Reduction (C/C_o)	Removal Time (hr)	
	Charcoal	Catalytic Oxidizer
0.5	3.36	12.1
0.1	11.2	40.2
0.01	22.4	80.4

Thus, the TCCS seems to be inadequate to respond to an accidental release of an acutely toxic chemical. NASA intends to develop measures to respond to sudden releases, but hardware for this purpose is not part of the baseline station configuration. Consideration also should be given to the alternative of increasing the capacity of the TCCS as a means of responding to accidental releases. The design criteria should take into account category 2 sources and the 1- and 24-hr SMACs of contaminants released from these sources.

Another concern regarding the design of the ARS is the relative locations of the dehumidification unit and the TCCS. Condensate from the cabin atmosphere, although vented overboard during the man-tended phase, will be used after treatment as potable water in the permanently manned configuration. Air filters for removal of particulate matter will be located upstream of the dehumidifier. Nevertheless, the cabin air condensate is expected to contain numerous volatile organic compounds that are either present as vapor in the return air stream or vaporize from particulate material collected on the filters. Condensate from space-shuttle missions has been shown to contain a wide variety of organic contaminants.

Air Monitoring

Because of the high rate of air circulation, it is unlikely that personal (breathing zone) monitoring will yield much better estimates of exposure than area monitoring, with the possible exception of exposures experienced by those working very close to a contaminant source. For those personnel, the use of passive air sampling or biological monitoring or both should be considered.

General

The dynamic mass-balance model of contaminant emission and removal rates available at NASA should be used to predict the temporal contaminant concentration profiles associated with all reasonably probable emission scenarios. (Probable is defined by the Federal Aviation Administration as an event having a probability of at least 10^{-5}/flight-hr (FAA, 1988).) These scenarios should include low-level continuous emissions, periodic releases, and abrupt high-level releases. The concentration maximum and the 1-hr, 24-hr, and 180-day time-weighted averages could then be calculated from the hypothetical concentration-time profiles and compared with the SMAC values. This model should include humans as both sources and sinks.

Results of these tests could be used to identify the contaminants of greatest concern and to take appropriate precautions. Continuous monitoring, alarm systems, and emergency response procedures would be required if exceeding a short-term limit (1-hr SMAC) is reasonably probable. Periodic air monitoring or biological monitoring or both as well as deliberate response procedures would be sufficient for most other contaminants. In addition, the use of the model in this manner would help identify gaps in the current understanding of system performance.

The overall plan for monitoring air contaminants, biological monitoring of the crew, and complying with the SMAC values should be developed by a team including industrial hygienists, physicians, toxicologists, behavioral toxicologists, and health physicists, as well as chemists and engineers.

5 Physiological Changes During Spaceflight

The general physiological effects of exposure to microgravity are better understood but much remains to be learned about the quantitative impact of physiological changes on susceptibility to chemicals. Because a correlation appears to exist between the duration of exposure and the frequency and intensity of the effects, the issues of duration of the spaceflights and of the microgravity environment remain the critical variables in attempting to establish safe levels of exposure to potentially toxic chemicals. Risk assessment of chemicals begins with biological studies conducted in animals and humans on earth. It is not clear how these data can be generalized to the space environment. A variety of assumptions will be necessary. The quality of these assumptions can be substantially improved as more data from astronauts and cosmonauts are obtained. To date, of the effects observed in astronauts, space motion sickness and decreased immune function seem to be the effects having the most direct relevance to crew performance during spaceflights.

SPACE MOTION SICKNESS

Spatial disorientation and space motion sickness (SMS) occur early in spaceflight. SMS has been described since the Apollo 9 flight in the U.S. manned spaceflight program. The incidence of SMS among crew members has ranged from about 35% during Apollo to 60% during Skylab. During the four orbital flight tests of the shuttle, four of eight crew members described symptoms of SMS.

SMS is a persistent operational medical problem and has been called the most clinically significant medical phenomenon during the first several days of spaceflight. SMS is characterized by increased sensitivity to motion and head movements, headache, malaise, lethargy, stomach awareness, loss of appetite, nausea, and episodic vomiting.

Onset of symptoms usually occurs in the first 15 min to 6 hr of spaceflight, although delayed symptoms have been reported up to 48 hr into the flight. Neural adaptive mechanisms respond, and within a few days, the symptoms disappear. These mechanisms are not well understood and further research is required. Symptom severity peaks during the first 2 days of flight, and recovery is usually complete by the third or fourth day. Results from the Soviet Salyut 6 flights show that symptom resolution occurred after 3–7 days in space, but one crew member reported having symptoms of SMS for 14 days.

The high incidence, severity, and duration of SMS have limited some early flight crew activities. Extravehicular activity is now scheduled after the third flight day to allow symptomatic crew members to recover from SMS prior to extravehicular activity. A minimum-duration flight lasts at least three flight days to ensure that the crew has recovered sufficiently to permit all entry and landing activities. Until successful countermeasures or accurate preflight predictive tests can be developed, SMS will continue to have an impact on crew activities during the relatively short 5- to 10-day flights of the shuttle. The basic mechanisms of SMS should be defined so that drug therapy or other techniques for reducing SMS can be developed.

BONE AND MINERAL METABOLISM

A variety of studies of human beings during long-term bed rest, of humans in space, and of rats in space have shown that prolonged inactivity and weightlessness result in both significant and continuing losses of calcium from the skeleton and nitrogen from muscle and decreases in mass of both rats and humans (NRC, 1988b; Whedon, 1984). These changes were consistent but quite different in degree from subject to subject. In the longest bed-rest studies (7 months) and in the longest orbital spaceflight during which metabolic measurements were made (3 months), the rate of calcium loss was as great at the end of the studies as it was soon after the start. In the severe paralysis of poliomyelitis, calcium losses led to x-ray-visible osteoporosis in the bones of the lower extremities as early as 3 months after paralysis. While the overall rate of calcium loss in Skylab astronauts was 0.4% of total body calcium per month, the loss was estimated to be 10 times greater in the lower extremities than in the rest of the body (based on bed-rest studies of calcium losses by metabolic balance compared with decrease in bone calcium density). That loss could lead in 8 months of flight to a decrease in bone density in the legs similar to that noted in paralytic poliomyelitis. In longer flights, if mineral loss were to continue at a similar rate, the bones of the legs might fracture during physical work in as little as 9–12 months, especially at gravities approaching 1 g. Studies of immobilized rabbits showed marked decrease in strength of tendons and ligaments after only 1 month. Thus, strains, sprains, and even ligament tears may be more likely to occur and may occur earlier than bone fractures (Whedon, 1984). These risks are not likely to affect crew performance during flights, but they are serious considerations on return to earth after long flights or after landing on another planet, such as Mars.

The cellular mechanisms of mineral loss are unknown. Excess excretion of calcium associated with increased hydroxyproline in the urine in humans is indicative of increased bone resorption. Histological examination of the bones of the rats on Cosmos showed suppressed bone formation; it is difficult, however, to apply these results directly to humans because of differences in rat bone physiology.

In more recent research, bed-rest studies under NASA sponsorship have been continued in search of countermeasures that could be applied to astronauts in space to suppress or prevent calcium loss (NRC, 1988b; Schneider and McDonald, 1984). All mechanical procedures tested thus far have been ineffective including exercise with the Exer Genie pulley apparatus, static and intermittent compression of the length of the body from shoulders to feet, static and intermittent lower-body negative pressure, and impact loading up to 36 lb to the bottom of the heel 40 times per minute for up to 8 hr daily. Correlative observations have indicated that a procedure would have to be derived for use in flight that would provide an equivalent force on the skeleton of 4 hr of walking per day. A high calcium and phosphorus diet reduced calcium loss for up to 90 days only; thereafter, increasing fecal excretion of calcium rendered a negative calcium balance and continued to do so for the remaining 17 weeks of the study. Supplementary dietary phosphate alone had no lasting beneficial effect to prevent calcium loss during bed rest. Salmon calcitonin at 100 MRC (Medical Research Council) units daily also did not prevent calcium loss. Promise in these countermeasure studies has come from bisphosphonate compounds, such as disodium etidronate, that bind to bone crystal and tend to inhibit bone resorption (Schneider and McDonald, 1984). Countermeasure studies are continuing particularly on the bisphosphonate leads.

At the same time, with support from the National Institutes of Health and other sources, various studies are being conducted on the basic mechanisms of the effects of mechanical forces on bone dynamics and development. Such studies may give insight into the bone-loss problem in space. Conversely, development of effective countermeasures to bone loss in space may contribute to improved therapy or management of osteoporosis, which is characterized by gradually decreasing bone mass and strength and is the most prevalent clinical disorder of bone.

As another mineral metabolic effect, hypercalciuria associated with loss of mineral from bone in spaceflight might increase the potential for stone formation in the urinary tract (NRC, 1988b). Although 75-80% of renal stones contain calcium, the possibility of stone formation appears to depend not only on increased urinary concentration of calcium but also on other factors such as urinary pH, concentrations of inorganic elements (magnesium, potassium, and phosphorus), and concentrations of organic compounds (uric acid, citrate, and oxalate). Bed-rest studies have shown a slight rise in urinary pH and a lack of change in urinary citrate, which in ambulatory states rises with increases in urinary calcium (Deitrick et al., 1948). Both of these factors, if also noted in spaceflight, would favor decreased solubility of calcium salts. These considerations suggest that research ought to be continued on urinary-tract stone formation in relation to microgravity as a significant possibility during long spaceflight. The likelihood of such an occurrence may be small, especially if care is taken to maintain abundant urine volumes; nevertheless, such stone formation might be catastrophic to health and function for the astronaut involved and thus to success of the particular flight.

Potential toxicants having an adverse effect on bone that might be introduced into the environment aboard the space station are aluminum and fluoride. Aluminum, a component of antiperspirants, could be inhaled or absorbed through the skin and, if used in continuing substantial quantities, could be deposited in bone and inhibit bone formation. Fluoride has been under consideration as a possible inhibitor of bone resorption in weightlessness, but it should not be used indiscriminately, because at only moderately high concentrations, its incorporation into bone makes bones brittle and at increased risk of fracture. Other cations and anions in the environment could have adverse effects on calcium and bone metabolism.

MUSCLE METABOLISM

After a few days of exposure to microgravity, the urinary excretion of nitrogen compounds increases and muscle atrophy begins. These effects may compromise the ability of astronauts to do their jobs. They may not be able to withstand the stress of 1 g upon return to earth; the continued excretion of nitrogen may have deleterious hormonal and nutritional effects. Exercise, diet, or drugs may ameliorate these effects, but a fully effective treatment is not likely to be developed until impaired muscle function in prolonged microgravity is better understood.

The increased urinary excretion of nitrogen by astronauts in Skylab reflected muscle loss such as that observed during bed rest, but the excretion was variable and generally greater than that seen during bed rest. Most of the atrophy occurred in antigravity muscles, which are no longer load-bearing.

In all nine Skylab astronauts, the high level of nitrogen excretion continued unabated for the duration of the flight (up to 84 days) (Whedon et al., 1974). This response indicates a serious malfunction not likely to reach a new steady state until an extreme degree of atrophy is reached. The nitrogen loss was accompanied by losses of 15-30% of muscle mass and strength in the lower extremities. The considerable exercise activity of the astronauts in Skylab 4 resulted in somewhat lesser losses of muscle mass and strength than on earlier flights but was obviously not fully protective.

Although the mechanism of the process of atrophy remains unknown, certain aspects have become evident (NRC, 1988b). Muscle atrophy is accompanied by decreased synthesis of muscle protein and by some degree of increased degradation. As shown in rats that are suspended (hind limb unloaded), loading and stretching of otherwise inactive leg muscles prevented muscle atrophy and stimulated protein synthesis; the addition of electrical stimulation increased protein synthesis markedly. As shown in muscle cultures, stretching stimulates protein synthesis.

The uncertain value of physical exercise for suppressing muscle atrophy during spaceflight has been noted previously; no controlled studies of exercise in flight have been attempted.

CARDIOVASCULAR FUNCTION AND BODY FLUID CHANGES

In the Space Science Board's publication *Life Beyond the Earth's Environment* (NRC, 1979), the Cardiovascular Panel summarized their review of studies up to that date in this area of space physiology as follows:

> The experience hitherto gained from manned spaceflight demonstrates that the cardiovascular system can adapt promptly to weightlessness and that man can maintain an excellent functional capacity in space for prolonged periods of time. However, an impressive body of data indicates that sudden exposure to 0 g is associated with a rapid shift of a considerable amount of interstitial fluid and blood from the lower toward the cephalad parts of the body. Most of the translocated fluid is accommodated in the intrathoracic compartment, distending its vascular structures and presumably inducing significant changes in central hemodynamics. The increase of the intrathoracic blood volume is apparently "interpreted" by the body as a "total body" intravascular volume expansion and elicits compensatory mechanisms (i.e., natriuresis and diuresis), which reduce total blood volume. The lack of any impairment in inflight physical work (aerobic) capacity indicates that the contracted blood volume is an appropriate adaptation to the zero environment. However, this adaptation becomes inappropriate upon return to normal gravity. Postflight circulatory studies demonstrate decreased orthostatic tolerance, decreased physical work capacity, and lowered exercise stroke volume and cardiac output in the sitting position. All of these postflight phenomena are consistent with a state of relative hypovolemia.

The fluid shift from the legs into the thorax as seen in Skylab crew members occurred within the first few days of flight and amounted to more than 2 L of extravascular fluid, as determined by a series of accurately located limb-girth measurements (Thornton et al., 1974). Related Skylab studies showed an apparent increase in leg venous compliance during the first 2 weeks of flight and a later decrease thought to be primarily related to the decrease in leg musculature (Thornton and Hoffler, 1974). In addition to the apparent decrease in plasma volume in weightlessness, red-blood-cell volume in Skylab crew members decreased by a mean of 11%, or 232 mL. Although some hemolysis occurred in Gemini astronauts, neither hemolysis nor hemorrhage occurred in Skylab members, and current interpretation is that splenic entrapment associated with slight inhibition of bone-marrow red-blood-cell formation occurred, as indicated by an observed decrease in reticulocyte counts (Johnson et al., 1974).

IMMUNE SYSTEM

Although reports to date are conflicting, some indicate that a microgravity environment may alter the immune system's function. Cogoli et al. (1980) reported that cultures of human lymphocytes subjected to microgravity responded to concanavalin A, a lymphocyte-stimulating agent, 97% less frequently than ground-based controls did. Studies of the astronauts of the first four space-travel-system flights revealed that the lymphocyte responses to phytohemagglutinin, another lymphocyte-stimulating agent, were reduced by 18–61% of normal following spaceflight (Taylor and Dardano, 1983). Stress has been suggested as the cause of these changes, but that cause has not been established and should be studied further.

In an unmanned Russian spaceflight, the weights of lymph nodes and spleens of rats flown for 22 days were reported to be markedly reduced, compared with those of controls on earth, because of a marked decrease of lymphocytes in these organs. The effects were found to be reversible since the organs returned to normal 27 days postflight (Durnova et al., 1977). In another study, Mandel and Balish (1977) studied rats subjected to a 20-day flight aboard the unmanned U.S.S.R.–Cosmos 7820. They immunized groups of rats with formalin-killed *Listeria monocytogenes* 5 days before flight and compared animals exposed to space conditions with 1-g controls. They concluded that no deterioration of the acquired cell-mediated immunity to *L. monocytogenes* could be detected in flown rats.

NUTRITION

Prior to the start of the spaceflight program, there was speculation that decreased effort of movement in weightlessness would result in diminished caloric requirements compared with those on earth. Diets were actually planned, however, at caloric levels close to those needed for normal activity on earth. In practice this procedure has worked reasonably well. In the 1- to 3-month flights of Skylab, modest loss of body weight occurred, associated with body-fluid shifts and losses in muscle mass, as astronauts consumed 2,400–2,800 calories per day. Clearly, caloric requirements were not lessened in space (NRC, 1988b).

In the past, many athletes and astronauts have been convinced that high intake of protein builds muscle and strength. However, the physiological evidence indicates that protein is increased in muscle only when needed for the muscle hypertrophy required by continuing physical activity; excess calories of any kind are converted to and stored in the body as fat. In addition, numerous studies unrelated to space have indicated that increasing the protein intake increases the urinary excretion of calcium. The level of protein in the diets of astronauts, therefore, should be reconsidered because of its potential relationship to urinary tract stone formation and, possibly, loss of minerals from the skeleton. The high phosphate content of meat may partially protect against the effect of high protein intake increasing urinary calcium. At the same time, the negative nitrogen balance associated with muscle atrophy in weightlessness should not be accentuated by encouraging too low a protein intake. Since negative nitrogen balance in space has occurred at daily protein intakes of 85-95 g, the recommended intake should not fall below this level (NRC, 1988b).

Carbohydrates should be of special concern because of their effects on behavior. Abundant evidence supports the view that any dietary carbohydrate that elicits the secretion of insulin can increase the synthesis and release of the brain neurotransmitter serotonin unless consumed with adequate amounts of protein. This substance makes people drowsy and interferes with optimal performance. If this relationship is not recognized, menus and the time of consumption of particular items—especially snacks—might not be appropriate for the tasks required, particularly if they are complex and prolonged. It is possible that other food constituents will be found that affect behavior, mood, and cognition (NRC, 1988b).

Among the countermeasures tested by NASA have been high intakes of calcium and phosphorus by bed-rest subjects. The study showed that this procedure maintained calcium intake and excretion level in balance for up to 3 months, after which the gradually rising fecal excretion of calcium caused a negative calcium balance. Hence, there is no basis at this time for recommending calcium intake during spaceflight at a higher level than 1,000 mg/day. It is obvious, however, that a low intake of calcium favors loss of this mineral from the skeleton, adding to the deleterious effect of microgravity. Therefore, the recommendation of the Life Sciences Task Group of the Space Science Board is 1,000 mg/day (NRC, 1988b).

Bed-rest studies of the effects of high intake of phosphorus showed some suppression of the tendency of urinary calcium to elevate, but overall phosphorus intake manipulation was ineffective because of gradually increasing fecal calcium excretion. Furthermore, the possible deleterious effect of phosphorus intake higher than an approximate calcium/phosphorus ratio of 1:1.8 must be remembered. Too high an intake of phosphorus will exert some binding effect on calcium in the intestine and tend to inhibit calcium absorption.

Since no studies have been done on the effects of spaceflight on the metabolism of any of the trace elements, no comment can be made other than that care should be taken that space diets contain trace elements in the amounts recommended in the U.S. Recommended Dietary Allowances (RDAs).

The important vitamin in long spaceflights is vitamin D. Enclosure in a space vehicle will prevent the normal conversion in the skin of the vitamin D precursor to vitamin D. Conversion is normally accomplished by exposure of the face and arms to as little as 20-30 min of sunlight a day. Since vitamin D is essential for facilitating calcium absorption from the intestine, as well as other calcium-related effects in kidney and bone, this vitamin will need to be supplied to space travelers. However, amounts should not exceed 800-1,000 IU/day (Holick, 1986). The RDA of vitamin D is 200 IU (5 μg) (NRC, 1989a); however, there is evidence that in the absence of any exposure to sunlight, the RDA for healthy young adults, such as astronauts, is closer to 600 IU (15 μg) (Holick, 1987).

The lighting environment, including the spectral distribution and intensity of the lighting, needs to be carefully engineered for the space station. Adding a small component of UVB (290-320 nm) radiation in an area where astronauts exercise or eat would promote vitamin D in the skin. The intensity of the lighting also should be evaluated to help maintain the biological clock and decrease the incidence of "seasonal affective disorder syndrome."

Other vitamins are not as critical since adequate amounts will be taken in the diet provided it is balanced and the vitamins are not degraded by the methods of food preservation in use. It has become customary, however, to provide astronauts with daily vitamin supplements at RDA levels, which is a reasonable procedure.

In the early days of planning for manned spaceflight, many thought that diets should be low in residue so that bowel movements would be small and infrequent. However, bowel function in microgravity, especially in longer flights, was observed to be essentially normal. Hence, diets should be normal in residue, and adequate bulk should be available to afford relatively easy passage of stools once or twice a day.

With regard to research and development at the practical level, the acceptability of various currently available packaged, canned, freeze-dried, or heat-stable food items should be evaluated for spaceflights extending many months to years. Because the capacity to carry and store frozen food items is likely to be limited in extremely long flights, research in space-food technology should be revived in planning for the space-station era. To date, nutrition investigations (unrelated to space) suggest that individuals do not crave a continuous variety of foods, but rather they tend to select from a small range or limited number of foods over a period of months, and these periods continue throughout life. Reduction in the total list of available food items should simplify both the strategy of storage of multiple food packages in a long-flying spacecraft and the ability of travelers to retrieve desired items with a minimum of difficulty and time. The space station will need to provide for testing of currently available and newly formulated items for long-term durability and acceptability.

6 Establishment of Spacecraft Maximum Allowable Concentrations

DESCRIPTION OF SMACs

Spacecraft maximum allowable concentrations are used as guidance for chemical exposure, either during normal operations of the spacecraft or during emergency situations. A 1- or 24-hr SMAC is defined as a concentration of a substance in air (such as a gas, vapor, or aerosol) that may be acceptable for the performance of specific tasks during emergency conditions lasting for periods of less than 1 hr or less than 24 hr. The effects of an exposure at a 1- or 24-hr SMAC could include reversible effects that do not impair judgment and do not interfere with proper responses to the emergency. The kinds of emergency exposures anticipated could result from events such as fires, spills, or line breaks.

The 1- and 24-hr SMACs are acceptable only in an emergency when some risks or some discomfort must be endured to prevent greater risks (such as fire, explosion, or massive release). Even in an emergency, exposure should be limited to a defined short period. Exposure at the 1- and 24-hr SMACs might produce such effects as increased respiratory rate from increased carbon dioxide, headache or mild central nervous system effects from carbon monoxide, and respiratory tract or eye irritation from ammonia or sulfur dioxide. The 1- and 24-hr SMACs are exposure levels that should not cause serious or permanent effects. While minor reduction in performance is permissible, it should not be so much as to prevent proper responses to the emergency (such as shutting off a valve, closing a hatch, removing a source of heat or ignition, or using a fire extinguisher). For example, in normal work conditions, a degree of upper respiratory tract irritation or eye irritation causing discomfort would not be considered acceptable; during an emergency, such an effect would be acceptable if it did not cause irreversible harm or seriously affect judgment or performance.

SMACs for up to 180 days are concentrations designed to avoid adverse health effects, either immediate or delayed, and to avoid degradation in performance of crew after continuous exposure. In contrast to 1- or 24-hr SMACs, which are intended to guide exposures during emergencies (exposures that, although not acceptable under normal operating conditions, should not cause serious or permanent effects), 180-day SMACs are intended to provide guidance for operations lasting up to 180 days in an environment like that of a space station. Accumulation, detoxification, excretion, and repair are important in determining 180-day SMACs. If a material is cumulative in its effects, its 180-day SMAC must take that into account. Neuropathological regeneration or repair of toxic injuries occurs more readily in intermittent exposures than in continuous exposure to constant toxic insult. Therefore, repair is important in the recommendation of 180-day SMACs.

SOURCES OF DATA FOR DEVELOPING SMACs

Various types of evidence should be assessed in establishing SMAC values. These include information from (1) chemical-physical characterizations of the potential toxicant, (2) in vitro toxicity studies, (3) animal toxicity studies, (4) human clinical studies, and (5) epidemiological studies.

Chemical-Physical Characteristics of Toxicant

The chemical and physical characteristics of a chemical provide valuable information on the dosimetry of the compound within the body and on the likely toxic effects. For example, size and water solubility of inhaled particles strongly influence where the material deposits in the respiratory tract. Likewise, lipophilicity influences absorption of the material where it accumulates and how long it remains in the body. Structure-activity relationships may allow estimation of the toxic potential of new compounds based on the known toxicities of well-investigated structurally related compounds. However, additional uncertainty (safety) factors must be applied to arrive at safe levels for those congeners that have no dose-response data from intact animals.

In Vitro Toxicity Studies

Important data can be obtained from studies that investigate adverse effects of chemicals on cellular or subcellular systems in vitro. Systems in which toxicity data have been collected include isolated organ systems (e.g., isolated perfused livers and lungs), single-cell organisms including bacteria, cells isolated from specific organs of multicellular organisms and maintained under defined conditions (e.g., isolated hepatocytes and bone-marrow colony-forming units), functional units derived from whole cells (e.g., organized subcellular particles including nuclei and mitochondria), breakdown products of cellular disruption (e.g., microsomes and submitochondrial particles), isolated or reconstituted enzyme systems, and specific macromolecules (e.g., proteins and nucleic acids).

Data on inhibition of specific physiological functions, pathological outcomes of exposure, genetic damage, changes in xenobiotic metabolism, or changes in levels or quality of cellular components can contribute to these evaluations. In vitro studies can help both to describe the effects of chemicals and to provide information on the mechanism of action of chemicals.

In vitro systems are used on the assumption that the effects observed present a reasonable model for humans. In the current context, the additional caveat that they should reflect the response of humans in space must be added. Therefore, use of older data and plans to collect new data should pay heed to the need for surrogate modeling relevant to the astronaut.

Animal Toxicity Studies

The data necessary to evaluate the relationship between exposure to a pollutant and its effects on a population are frequently not available from human experience. For many air pollutants, studies in animals have provided the only useful data. Ideally, the data should be derived from at least two species and by the inhalation route.

Inhalation experiments with animals provide a basis for estimating possible effects in humans and the concentrations at which these effects occur. They are useful in the identification of adaptations that may occur following repeated exposure. They permit the testing of hypotheses about the mechanism of the toxic action of pollutants.

They offer a good opportunity to explore interactions between pollutants and other factors that may affect toxicity.

Data from skin absorption, ingestion, and parenteral studies are also potentially useful. Since eye irritation can be debilitating, eye-irritancy testing of substances found in space-station air also is needed. The usefulness of animal data depends in part on the species used. Relevance to humans may be limited in the absence of information on target organs and pharmacokinetics in both animals and humans.

The better animal studies will report the following:

- The most sensitive target organ(s) or body system(s) affected by exposure to the contaminant in question.
- The nature of the effect on the target organ(s).
- Data to establish dose-response relationships for the target organs(s)—from no effect to severe effects. (The distinction between exposure and dose needs to be made.)
- The rate of recovery from reversible effects, if any.
- The nature and severity of injury for effects that are not reversible.
- Cumulative effects, if any, such as neurotoxicity and cancer.
- Pharmacokinetic data for comparison with data obtained from humans.
- The effects of interaction, if any, of the toxicant with other air pollutants (or exposure conditions) and the minimum concentrations at which the interaction appears to occur.
- Techniques used to assure quality and avoid bias.

Classic toxicity studies employ normal animals. It may be necessary to develop animal models with features similar to the physiological state of the astronaut in prolonged spaceflight. Although such models are approximations to the human condition, they should provide better information than studies on unaltered normal animals. Thus, animals flown in the space shuttle or the space station are likely to be more appropriate surrogates for humans. For example, rats flown aboard Cosmos 1887 showed altered hepatic function. These rats also demonstrated skeletal muscle weakness resulting from muscle fiber atrophy and segmental necrosis. Studies of the myocardium showed evidence of atrophy.

In addition to the microgravity of space, animals can also be exposed to launch and reentry gravity forces, noise, and vibration.

Clinical Studies

In establishing SMACs for chemicals, dose-response data from human exposure are most desirable, and such data should be used whenever possible. Experimental human studies can be designed to provide useful information on dose-response relationships. Ethical concerns limit these studies to pollutants that are anticipated to have no residual effects consequent to the experimental exposure and to short-term studies. Data from inhalation exposures are most useful here, because inhalation is the most likely route of exposure.

Epidemiological Observations

Human toxicity data frequently are obtained from epidemiological studies of long-term industrial exposures as well as short-term exposures usually to high levels of toxicants following accidents. These data sometimes provide a basis for estimating a dose-response relationship.

Epidemiological studies have contributed to our knowledge of the health effects of many airborne chemical hazards, for example, radon daughters (Whittemore and McMillan, 1983) and vinyl chloride (Waxweiler et al., 1976). Studies of environmental and occupational exposure can assess acute effects of short-term exposure such as myocardial infarction after exposure to methylene chloride (Stewart and Hake, 1976) or effects of long-term exposures such as cardiovascular disease associated with long-term exposure to carbon monoxide (Stern et al., 1988). One limitation of most epidemiological studies is the limited information that is available about past exposure (Checkoway et al., 1989). Studies that include estimates of past exposure based on available historical exposure records, either of the work environment or personal samples, tend to be more useful than those based on years of employment (Rinsky, 1989). Studies that involve assessment of exposure that occurs while the cohort is followed are likely to provide more reliable information on exposure because the sampling scheme can be devised as part of the research plan rather than relying on available data collected for other purposes, such as assessing compliance with exposure regulations (Smith, 1987). However, an appealing aspect of retrospective studies is that the exposure, the interval between exposure and onset of disease, and the onset of disease will have occurred by the time the study is conducted; in prospective studies, time must pass for any responses to occur. Epidemiological studies also vary in the accuracy and precision of the health outcome measured. Some of these outcomes must rely on available information such as death certificates, which may be incorrect at times (Percy et al., 1981), while others rely on clinical testing, pathological reports, or early preclinical markers of pathology. Despite these limitations, if the populations studied are large enough, had substantial exposure, and had sufficient interval between exposure and study to allow for the expression of disease, epidemiological studies have the major advantage of considering the effects of exposure in humans. Epidemiological studies can often provide the basis for establishing a permissible concentration for human exposure.

Epidemiological outcomes often are reported in terms of relative risk. Relative risk is a ratio of the rate of outcome of either disease or disability in the exposed population to that in the nonexposed population. Relative risk is not a measurement of risk. The relative risk for a rare disease can be equivalent to that of a common disease and lead to substantially less total risk. To establish a level of acceptable exposure, information on relative risk usually must be reformulated to understand the relationship of relative risk and risk of morbidity or mortality. The level of risk that is acceptable is a matter of policy, not epidemiology.

TYPES OF DATA USED IN RECOMMENDING SMACs

The preceding section described the sources of data for use in establishing SMACs. In this section, the two types of data obtained from these sources and their importance in establishing SMACs are described. The types are (1) dosimetry and (2) toxicity end points.

Dosimetry

Deposition of Particles, Gases, and Vapors in the Respiratory Tract Under Microgravity Conditions

Aerosols deposit in the respiratory tract of people primarily by the processes of sedimentation, impaction, interception, and diffusion. Sedimentation is due to the gravitational force acting on the particles, and the particles may settle and deposit on the lower surfaces of airways. Impaction is due to the inability of individual particles to follow the curvature of air streamlines because of inertia, and the particles may hit and stick on the walls of airways. Diffusion is due to random (Brownian) motion of small particles caused by interaction with air molecules, which may cause a particle to move across the air streamlines and deposit on contact with the airway wall. In a microgravity environment, sedimentation will not be a mechanism of deposition, and deposition will be mainly due to inertial impaction and diffusion.

For small particles (<0.2 μm), deposition is dominated by the diffusion mechanism, which is independent of gravity. Gas and vapors also come in contact with airway walls due to Brownian motion. Thus, the deposition of small particles, gases, and vapors in the respiratory tract will not be significantly affected by microgravity conditions. As the particle size increases (1-5 μm range), sedimentation becomes an increasingly important mechanism of deposition under normal gravity forces. Therefore, fractional deposition of particles >1 μm in the respiratory tract will be significantly less under microgravity conditions. The pattern of deposition also will shift so that proportionally more of the large particles will deposit in the tracheobronchial region, as opposed to the alveolar region of the respiratory tract. In addition, sedimentation is important for particles >5 μm because it greatly affects the persistence of these particles in air. The concentration of particles from 5 to 100 μm and greater inside the spacecraft will be much higher under conditions of microgravity. These large particles are likely to be very irritating to the eye and the respiratory tract. Transmission of infections can increase because large droplets of saliva and respiratory secretions remain suspended in the atmosphere for a longer time under microgravity conditions than under normal conditions. Changes in total deposition, regional deposition, and concentration of large particles in air could influence the potential health effects of particles in the space environment.

Thus, the absence of gravity makes a significant difference in particle deposition only when particles range in size from 0.5 to 2 μm. Diffusion will account for most deposition for vapors and for particles smaller than 0.5 μm, and impaction will be responsible for most particle deposition for particles larger than about 2 μm in aerodynamic diameter.

Pharmacokinetics and Metabolism

The evaluation of the health effects of any chemical in a given environment is often improved by an understanding of its physiological disposition in the body (pharmacokinetics and metabolism). Metabolism of a chemical may lead to detoxication or may result in metabolic activation leading to toxic effects. Since metabolic events are usually enzyme mediated, they are driven by the concentration of substrate available for the reaction. The concentration of substrate is a result of the level of exposure and the pharmacokinetics of the chemical. Modern pharmacokinetic studies are increasingly aimed at developing dispositional models, which compare input of chemical with outflow for the whole system as well as for individual organs. The space station is a system that can be modeled for any given chemical. The space station is a closed system with limited capacity to clear the air of chemical vapors, while the crew contributes to the removal of some chemicals through sequestration and metabolism. Thus, physiologically based pharmacokinetic models of the disposition of inhaled materials should be useful in helping to assess the risk of disease from airborne toxicants in the space station.

The toxic metabolites of each chemical and factors that control the rate at which they form are important for SMACs. It is important to distinguish between the production of toxic metabolites in organs where toxicity is observed and metabolism in liver where most chemicals are metabolized. The relationship between metabolism in liver and toxic effects in either liver or other organs is an important factor because metabolism of chemicals in a liver already damaged by other exposures is likely to differ from metabolism in a normal liver. Furthermore, if toxic metabolites are moved from the liver to other organs, damaged livers may alter the production of toxic responses (Merrill et al., 1990).

To determine the possible biochemical effects of prolonged weightlessness on liver function, samples of livers from rats that were on board Cosmos 1887 were analyzed for a number of key enzymes involved in metabolism of compounds and xenobiotics (Merrill et al., 1990). They observed slightly lower than normal amounts of cytochrome P-450 in the livers of the flight group; this finding was similar to that of Spacelab 3 (Merrill et al., 1987). They also observed decreases in the enzymes aniline hydroxylase and ethylmorphine-N-demethylase. It is not known with certainty whether spaceflight alters the body's ability to metabolize drugs; however, these limited results suggest that a change in the array of cytochrome P-450 enzymes occurred in the livers of the animals. This family of enzymes is responsible for the metabolism of drugs, the metabolism of a number of natural steroid hormones and their metabolites, and the metabolism of several compounds and mediators critical for intracellular communication and regulations, e.g., prostaglandins and various growth factors. An important function of these compounds appears to be their role in the hematopoietic system. Therefore, the normal functioning of the cytochrome P-450 system could have important implications for space biomedicine.

Toxic metabolites have been observed to be highly reactive chemically and to react with nucleic acids, proteins, or lipids to alter normal biological function. These metabolites may induce alterations in DNA replication or the process of transcription. Attempts to repair damage to DNA may involve misrepair, leading to erroneous DNA replication or function, inhibition of protein synthesis if RNA is the target, or inhibition of the enzyme or other activity if proteins are the target. In addition to the formation of reactive intermediates that are metabolites of the chemicals, metabolic activity may give rise to species of active oxygen, which may damage nucleic acids or proteins or yield lipid peroxidation. The effects may range from target-organ toxicity to carcinogenesis.

Biological Markers

Recently, the concept of biological markers as indicators of exposure to polluted air has been investigated (NRC, 1989b). Biological markers within an exposed individual can indicate the degree of exposure to a pollutant, the initial structural, functional, or biochemical changes induced by exposure, and, eventually, the changes associated with adverse health effects.

Biological markers are indicators of change within an organism that link an exposure to polluted air to subsequent development of an adverse health effect. It is convenient to divide biological markers into three groups: (1) biological markers of exposure, (2) biological markers (early predictors) of effects of exposure, and (3) biological markers of susceptibility to effects of exposure.

Biological markers of exposure can be thought of as "footprints" that the chemical leaves behind after interaction with the body. Such markers contain the chemical itself or a metabolic fragment of the chemical and, thus, are generally chemical-specific. Examples of such markers are the chemical adducts formed between alkylating agents and macromolecules, such as nucleic acids or proteins, particularly blood proteins. Another example is the presence of a volatile chemical in exhaled air. This type of marker has been used by Wallace (1987) to assess exposure to benzene during filling of gasoline tanks in passenger cars, exposure to tetrachloroethylene in dry-cleaning shops, exposure to chloroform from contaminated hot shower water in homes, and exposure to volatile aromatic compounds in tobacco smoke. The breath of crew members of the space shuttle or the space station could be assayed for volatile chemicals at the end of a tour of duty to assess exposures during spaceflights; similar measurements are made on submariners on extended cruises (Knight et al., 1984, 1985). Physiologically based pharmacokinetic models (Ramsey and Andersen, 1984) have been used to relate biological markers of exposure to prior exposure conditions.

Biological markers of the effects of inhalation exposure can be any indication of a chemically induced disease. The biological markers of greatest interest are those that are early predictors of late-occurring effects. Such markers would be invaluable in assessing what levels of pollutants can be tolerated in the space station without causing irreversible deleterious health effects. For example, cell proliferation may result in clonal expansion of initiated cells (Swenberg, 1989). Thus, persistent cell proliferation could be a biological marker predictive of an increased incidence of late-developing neoplastic lesions. Few markers, however, have been validated as predictive of late-stage diseases.

The third type of biological marker, markers of increased susceptibility to the effects of exposure to airborne chemical pollutants, is a potentially important, useful tool. Such markers possibly could be used to predict which persons are more likely to be adversely affected by space-station exposure. For example, polymorphisms related to acetylation and DNA repair can be related to susceptibility to chemically induced tumors. Such indicators of susceptibility must be used cautiously, however, as they pose numerous moral and ethical problems.

The appearance of metabolites in excreta have been useful in the past and will be even more useful in the future as biological markers of exposure, effect, and susceptibility. Metabolites that are specific to the chemical will be most useful. When similar compounds related to normal diet are found, the interpretation of the results may be more complex. Other types of biological markers that are undergoing evaluation include evidence of chromosomal damage and covalent binding of reactive metabolites to DNA or proteins in circulating blood cells.

Toxicity End Points: Humans and Animals

Mortality

Short-term exposure to a pollutant at high concentrations may result in death, which may be immediate or delayed. Mortality is a widely used and important index in animal experiments and is also most useful in epidemiological studies as an index of serious health effects. All causes of death may be used as data for SMACs; however, certified causes of death in humans are inaccurate at times, making it more difficult to associate exposures with toxic outcomes.

Morbidity (Functional Impairment)

Pollutants also produce functional impairment. Most of the considerations that apply to mortality as useful data also apply to morbidity. It is usually much more difficult to obtain reliable information on morbidity than on mortality. Whereas statistics of mortality are collected routinely on whole populations, information on morbidity usually is not. Health surveys such as the Health Interview Survey (HIS) and Surveillance Epidemiology and End Results (SEER) (cancer data) collect information on medical effects. Often, studies must be initiated to answer the exposure-response questions rather than relying on information collected for other purposes.

Clinical Signs and Symptoms

Physical examination and paying attention to signs and symptoms can contribute to the biomedical and behavioral assessment of exposure. Eye irritation and tearing, for example, are among the most sensitive indicators of excessive exposure to oxidants. Complaints of discomfort by exposed individuals provide an important criterion on acceptability of pollutants. Although they may not be related to any direct health effects, complaints of discomfort need to be considered as an indication of exposure and as a factor in setting acceptable exposure levels.

Pulmonary Effects

The main portal of entry of air pollutants is the respiratory tract. Respiratory symptoms, allergic sensitization, or changes in lung function provide evidence of the impact of pollutants. Spirometry is used by far the most often of the large number of tests used to measure changes in lung function. Widely used measures of the bellows function of the lung include the volume of air that can be expelled after a full inspiration (the forced vital capacity); the volume that can be expelled in a measured time, usually 1 sec (the timed vital capacity or forced expiratory volume in 1 sec, FEV_1); the ratio of the forced expiratory capacity to the forced vital capacity (expressed as a percentage); lung compliance; and inspiratory and expiratory resistances. Other useful tests are ventilation rate, gas exchange, and blood flow. A recent report from the NRC (1989b) on pulmonary markers gives detailed information on the use of some of these measures.

Hepatic Effects

An excellent source for reviewing the adverse effects of chemicals on the liver is by Plaa (1991) in *Casarett and Doull's Toxicology*. The liver is the single most important organ concerned with protecting the body against the invasion of foreign, potentially toxic chemicals and, therefore, is among the organs most vulnerable to chemical insult. The liver avidly accumulates many chemicals, often after oral administration, leading to the so-called "first pass effect." However, accumulation of chemicals and toxic responses of the liver may also occur after inhalation or exposure via other routes.

Some chemicals, such as phosphorus, may be directly toxic to the liver. Chemicals passing through the hepatic circulation may be acted upon to yield detoxification products or may be activated metabolically to become more toxic chemical entities. Metabolic activation of many chemicals leads to liver damage. Mechanisms by which liver injury may occur include the accumulation of lipids, lipid peroxidation, covalent binding of reactive metabolites to critical cellular macromolecules, depletion of antioxidants such as glutathione leading to oxidative stress, interactive toxicological effects originating in Kupffer cells that lead to damage in hepatocytes, disruption of calcium compartmentation, and other aberrations that may lead to cell death. The effects of chemicals may include local or generalized lesions leading to cell death and necrosis, inflammation, fatty infiltration, cholestasis, cirrhosis, or carcinogenesis.

Underlying these injuries may be factors such as excessive alcohol intake, diabetes mellitus, or starvation that induce increases in the levels of cytochrome P-450IIE1, an enzyme involved in the metabolic activation of many halogenated hydrocarbons and other chemicals to toxicologically active or carcinogenic species. Some halogenated hydrocarbons, such as halothane, can cause immunological responses in liver following repeated exposure. Nutritional deficiencies may exacerbate potential liver injury caused by chemicals such as ethanol.

Liver damage may be detected by the appearance of jaundice, measurement of hepatic enzyme levels in circulating blood, use of radioisotope accumulation in the liver, use of various imaging techniques, or, more invasively, evaluation of liver samples obtained by biopsy.

Reproductive and Developmental Effects

The reproductive and developmental effects of exposure to spacecraft toxicants could be either overt or subtle, and the full complement of potential reproductive effects that could occur in the space station is not known. Adverse effects on factors such as reproductive capacity or behavior in offspring of returned astronauts could escape detection easily against the background of human variation. These factors need careful consideration in risk estimation. An additional factor is that components of experimental packages are yet to be designed, although many of the supplies that are likely to be on board are known or will be determined. Each component merits careful consideration, and some additional testing may be needed because reproductive and developmental toxicity data are frequently absent from toxicological data bases (NRC, 1984d).

The potential effects of spacecraft toxicants on reproductive outcomes could be diverse both in nature and in mechanism. Examples of potential adverse effects in humans can be derived in large part from animal studies. These effects include testicular atrophy (Melnick, 1984), transient reduction in the number of spermatozoa

(Hanley et al., 1984), and in utero developmental problems or postnatal abnormalities of offspring (Kimmel and Buelke-Sam, 1988). For the purposes of this report, two assumptions may be useful concerning reproductive and developmental toxicities: (1) in general, adverse effects on reproduction and in utero development are threshold phenomena, and outside of direct-acting mutagens, exposures below certain levels have not been associated with adverse outcomes; and (2) an excess of any contaminant may cause adverse outcomes.

Acute exposure to chemicals at levels that produce overt adult toxicity, defined as disruption of homeostasis (Skalko and Johnson, 1987), may be considered as also capable of interfering with reproduction and development. The assumption could perhaps be made that such severe toxicity scenarios would not occur, but if they did, adverse effects on reproduction also would be a concern. After the exposure had ended and normal homeostasis was established, the reproductive system or a conceptus could have transient or permanent sequelae that could be subtle and delayed in appearance and be unrecognized as attributable to space-station-derived exposures.

The agents that merit special attention are those that interfere with reproduction or development at exposure levels too low to be considered acutely toxic. The concept of target-organ toxicity can be used to focus attention on those chemicals whose most vulnerable target is reproductive function or in utero development. Examples of such chemicals are ethylene glycol monomethyl ether, which has one stage of spermatogenesis as its most vulnerable target, and thalidomide, which, although having exceptionally low toxicity for adults, has an as yet undiscovered aspect of in utero development as its most vulnerable target.

Details of test protocols for reproductive and developmental effects testing are readily available from the U.S. Environmental Protection Agency (EPA, 1985). Guidelines for developmental toxicity risk assessment (EPA, 1991) and proposed guidelines for reproductive risk assessment (EPA, 1988a,b) discuss interpretation of data in this area.

Neurobehavioral Effects

Some pollutants have such subtle effects on the central nervous system (CNS) that they are detectable only by special behavioral tests (McMillan, 1987). For the most part, these tests are designed to measure sensorimotor performance—e.g., speed, accuracy, and fine discrimination (NRC, 1984d). A large number of prevalent spacecraft contaminants represent a potential major hazard because of their capability to alter CNS function and impair the performance of complex tasks (Anger, 1984). There are compounds for which extrapolating industrial threshold limit values (TLVs) to establish SMACs is inappropriate particularly if the industrial standards are based on gross toxic effects of the compounds and involve discontinuous exposures. Subtle effects, such as performance impairment, seldomly have been considered in establishing industrial TLVs. Data on the effects of contaminants on human performance are rarely available.

There is broad agreement, however, that screening for neurobehavioral toxicity can be carried out effectively with laboratory animals by measuring motor activity, schedule-controlled behavior, and morphological change in the CNS (Holson et al., 1990). The automated measurement of motor-activity patterns provides a continuous noninvasive assessment of a pollutant's effects on a stable performance baseline over an extended time interval (Dews, 1953; Reiter, 1977; Reiter and MacPhail, 1979). Schedule-controlled behavior, based on procedures for programming performance antecedents and consequences, can be used to measure specific memory and learning functions as well as sensory thresholds and reaction times (McMillan and Leander, 1976). The measurement of morphological change in the CNS can reveal serious neurobiological toxicity produced by environmental contaminants (Spencer and Schaumburg, 1980). Such assessments, however, require in situ perfusion and the use of contemporary tissue preparation for examination by light microscopy and, preferably, electron microscopy.

Carcinogenicity

Since the first manned spaceflights almost 30 years ago, the consequences of exposure to potential carcinogens (chemicals and radiation) have been a concern. To date there is no evidence that malignant disease occurs more frequently in astronauts and cosmonauts than in a similar cohort of relatively young persons. However, only a limited number of astronauts have been exposed, and to find significant excess cancer among them would imply a large increase in rates. It is widely accepted that cancer induction is a multiphasic process with at least three broad stages: initiation, promotion, and progression. Clearly, exposure to potent mutagens and known initiators must be avoided in the space-station environment. Identification of cancer promoters is more difficult and their implication in human-cancer induction continues to be a source of controversy. Chemical contaminants in the air of the space station should be assessed by accepted methods for their tumor-promotion potential. In general, promotion appears to be a dose-dependent phenomenon. However, establishing thresholds for promoters poses a significant problem for toxicologists.

The carcinogenic potential of chemical mixtures in the space-station environment needs to be considered. The carcinogenicity of mixtures at low concentrations is frequently assumed to be additive.

Mutagenicity

Nonlethal mutations induced by chemical or physical agents accumulate with prolonged or repeated exposure conditions (long-term effects) such as those expected on the space station. In general, a gene mutation involves a molecular change within a single gene (point mutation); a chromosomal mutation involves blocks of genes and results from breakage (altered microscopic structure resulting in a chromosome structural aberration) or from nondisjunction (aneuploidy). Somatic-cell mutations may cause damage in the exposed individual (carcinogenesis), whereas germ-cell mutations may cause abnormalities in future generations.

Rapidly accumulating data show that mutations in single genes or chromosomes are directly related to cancer initiation, promotion, and progression (Adams and Cory, 1991; Solomon et al., 1991; Weinberg, 1991) and to heritable developmental defects that result in miscarriage or offspring with one or more abnormalities (McKusick, 1990; Borgaonkar, 1991; Rinchik, 1991; Pawson, 1991). This impressive evidence for the impact of single mutations in humans makes imperative the inclusion of mutagenic data in assessing the risks of space-station contaminants.

Several approaches are used to gather the information needed to determine whether genetic risks should be included in deriving SMACs for given chemicals:

1. *Mutagens expected on board*. A literature search on the mutagenicity of specific compounds would reveal whether sufficient dose

(concentration × exposure time)-response data are available for making risk estimates. Data from short-term mutagen assays can be helpful in establishing permissible emergency peak concentrations as well as maximum allowable concentrations during long-term exposure.

2. *Human cell studies.* The human is a good "filter" for chemicals in air. Three cell types—the peripheral lymphocyte, the erythrocyte, and sperm—are human cells readily accessible for analysis of gene and chromosomal mutations and represent somatic and germ cells. In addition to mutation, molecular biomarkers can be detected in these cell types. Such markers are indicative of damage to crucial molecules.

- *Lymphocytes* (T cells) are excellent for detecting chromosomal changes and have been widely used for in vivo and in vitro mutagen assays to examine structural and numerical aberrations, sister chromatid exchanges, micronuclei, unscheduled DNA synthesis, and single-strand breaks in DNA. Few data are available, however, on chromosomal mutation in lymphocytes of astronauts (Gooch and Berry, 1969; Lockhart, 1974, 1977), and in light of current analytical methods, the data are inconclusive. A fair amount of information exists, however, about the response of lymphocytes to the conditions of spaceflight. Analysis of blood from astronauts on the first 12 shuttle flights revealed a significant decrease in the number of circulating lymphocytes (NASA, 1989). The ability of lymphocytes to be activated (undergo DNA synthesis and cell division) by mitogens (foreign antigens) is greatly diminished also or, in some cases, obliterated, as first reported after the 1961-1969 Soyus-6, -7, and -8 flights (Konstantinova et al., 1973) and later confirmed after the Skylab and Spacelab DI flights (Kimzey, 1977). Although the number and function (activation capability) of lymphocytes returned to normal after the flights, there is concern about the consequences of long-term missions.

Experiments suggesting that those effects on lymphocytes are caused by microgravity have shown that exposure of lymphocytes to 1 g in flight (centrifuge) produced no changes when compared with lymphocytes at microgravity (in vivo and in vitro) (NASA, 1989). Hypergravity (10 g) increases the response of lymphocytes to mitogens by as much as 500% over the level observed at 1 g (Lorenzi et al., 1986). Physical stress associated with marathon running causes the same effects on lymphocytes as does the stress of spaceflight (Gmunder et al., 1988). Of special interest with respect to chemical effects are reports that exposure to nitrogen dioxide at concentrations as low as 1.0 ppm suppresses T- and B-lymphocyte responses (Fenters et al., 1973; Maigetter et al., 1976; Richters and Damji, 1988).

- *Erythrocytes* and T lymphocytes are useful for detection of induced gene mutations (Albertini et al., 1990). The end-point mutations in erythrocytes are mainly in hemoglobin and GPA (glycophorin A) genes and in T lymphocytes, HGPRT (hypoxanthine-guanine phosphoribosyl-transferase) and HLA (human leukocyte antigen) genes.

- *Human sperm* chromosomes can be examined by fusing them with hamster eggs; this method has been used to study the mutagenic effects of radiation and some chemicals (Brandriff and Gordon, 1990). Detection of gene mutation in sperm is also possible (Wyrobek et al., 1990). Analysis of sperm head shape and motility provides a measure of the damage to male germ cells (Wyrobek and Bruce, 1978).

Mutations in either male or female germ cells may lead to reproductive and developmental toxicity (Kay and Mattison, 1985; Goldsmith et al., 1984). If toxicity were to occur, the effects usually would be difficult to detect and quantify in a given person because of variations among individuals (e.g., sperm count), because of time and manner of expression (e.g., congenital defects) and especially because of similar effects occurring spontaneously in other offspring. These difficulties should not diminish concern about their occurrence.

The induction of mutation may occur at any stage of spermatogenesis and result in a mutant sperm appearing immediately or years later. In a female, all but the final stages of oogenesis are completed prenatally, so that at birth the ovary contains only primary oocytes. Expression of a mutation induced in a primary oocyte could be delayed for years—i.e., delayed until the follicle containing it matures to ovulation and the fertilized egg develops into a fetus or even adulthood.

- *Molecular biomarkers* that are primarily carcinogen-DNA or carcinogen-protein adducts provide an extremely sensitive test of cell interaction with chemical contaminants. Minute amounts of such markers can be detected in DNA, e.g., 1 adduct in 10^9-10^{10} nucleotides; protein adducts in body fluids (blood, urine, breast milk, and semen), exhaled air, and adipose tissue; and new proteins, such as those produced by activated oncogenes (mutated genes that cause cancer) (Perera et al., 1991).

Immunotoxicology

Immunotoxicology is a science that explores the effects of chemical agents and other harmful substances on the immune system. The immune system can be either stimulated or inhibited by xenobiotics. Excessive stimulation can result in hypersensitivity or autoimmunity, and suppression can be expressed by an increased susceptibility of the host to infectious or neoplastic disease. Xenobiotic-induced immune dysfunction has been well established for several chemicals in animals. In some cases, the immune system has been identified as the most sensitive target organ to detect the minimum toxic doses of a xenobiotic. Although one or more of the many compartments of the immune system may be significantly suppressed, this suppression may not be directly expressed biologically as an immune-mediated disease but rather as a potential risk due to the reduced ability of the host to resist natural and acquired diseases.

Animal models have been extremely valuable in identifying immunotoxic agents and in developing immune profiles, identifying mechanisms of action, and alerting humans to potential health risks associated with exposure to specific xenobiotics, either consumed as drugs or through environmental exposure. Many immunoassays have been validated in animals to detect drug and chemical-induced immunomodulation. Some of these bioassays are sensitive and predictive in assessing immune dysfunction. Many other immune assays that either are in the developmental stage or are not being used extensively are likely to have application in immunotoxicology. In general, these procedures will require additional testing and confirmation before they can be widely accepted as validated. Although many tests are available to screen for immunotoxicants, choice of an initial test that evaluates T-cell-dependent antibody response permits assessment of several compartments of the immune system concomitantly. These procedures are easily performed, quantitative, sensitive, and economical, and they routinely detect a large percentage of the known immunotoxic agents.

A variety of tests are available to assess humoral and cellular immunity as well as nonspecific resistance in humans. A series of tests can be performed in steps to assess immune competence in individuals who have been exposed to an immunotoxicant or potential immunotoxicant. Some of these procedures parallel animal studies and require validation prospectively in populations exposed to putative immunotoxicants and in control groups to ascertain their predictive value as immune-compromising agents. This aggressive approach will permit use of sensitive procedures for detection of immunomodulation in humans.

RISK ASSESSMENT

Noncarcinogenic Effects

Toxicological risk assessment for agents without the capacity to induce carcinogenic or mutagenic effects has traditionally been based on the concept that an adverse health effect will not occur below a certain level of exposure, even if exposure continues over a lifetime. The existence of a so-called "threshold" dose is supported by the fact that the toxicity of many agents is manifest only after the depletion of a known physiological reserve. In addition, the biological repair capacity of many organisms can accommodate a certain degree of damage by reversible toxic processes (Klaassen, 1986; Aldridge, 1986). Above the threshold dose, however, the homeostatic physiological processes that allow compensatory mechanisms to maintain normal biological function may be overwhelmed, leading to organ dysfunction. Thus, the objective of classical toxicological risk assessment is to establish a threshold dose below which adverse health effects are not expected to occur or are extremely unlikely.

The concept of a no-effect level was introduced by Lehman and Fitzhugh (1954), who proposed that an acceptable daily intake (ADI) could be calculated for contaminants in human food. This concept was endorsed by the Joint FAO/WHO (Food and Agricultural Organization and World Health Organization) Expert Committee on Food Additives in 1961 and subsequently adopted by the Joint FAO/WHO Meeting of Experts on Pesticide Residues in 1962 (cf. McColl, 1990). Formally, an ADI is defined by the relationship

$$ADI = NOEL/SF,$$

where NOEL is the no-observed-effect level in toxicological studies (the highest experimental dose at which effects are not observed) and SF is the safety factor that allows for variations in sensitivity to the test agent in humans as compared with experimental animals and for variations within the human population. These two sources of variation have often been accommodated through the use of a 10 × 10 = 100-fold safety factor reviewed by NRC's Food Protection Committee (NRC, 1970).

In 1977, the NRC's Safe Drinking Water Committee reviewed the methods that had evolved for establishing ADIs and made several important recommendations. First, the committee proposed that NOEL be expressed in milligrams per kilogram of body weight rather than milligrams per kilogram of diet to adjust for dietary consumption patterns. Second, the committee suggested reducing the traditional 100-fold safety factor to only 10-fold in the presence of dose-response data derived from human studies. Third, the committee proposed augmenting the traditional 100-fold safety factor to 1,000-fold in the absence of adequate toxicity data (NRC, 1977).

Although the use of safety factors is now accepted practice in establishing exposures for noncarcinogenic effects, the NOEL/SF approach is subject to certain limitations (Munro and Krewski, 1981). Since ADI is only an estimate of the population's threshold dose, absolute assurance of safety is not provided (Crump, 1984a). Smaller experiments tend to yield larger NOELs, and hence larger ADIs, than larger, more sensitive experiments (Mantel and Schneiderman, 1975). Safety factors of 10-fold that are used to account for both inter- and intra-species variation in sensitivity cannot be guaranteed to provide adequate protection in all cases. For these reasons, ADI should not be viewed as possessing a high degree of mathematical precision but should be viewed as a guide to human exposure levels that are not expected to present serious health risks.

In 1988, EPA recommended using the term "uncertainty factor" (UF) rather than safety factor in recognition of the uncertainty associated with ADI and relabeled ADI as a reference dose (RfD) (Barnes and Dourson, 1988). EPA also introduced an additional modifying factor (MF) to account for specific scientific uncertainties in the experimental data used to establish RfD. The no-observed-adverse-effect level (NOAEL) is defined as the highest experimental dose at which no statistically significant increase in the occurrence of adverse effects is observed beyond that exhibited under control conditions. The RfD is determined using the relationship

$$RfD = NOAEL/(UF \times MF).$$

In the present context, adverse effect is defined as any effect that contributes to the functional impairment of an organism or that reduces the ability of the organism to respond to additional challenges (Dourson, 1986). When the data do not demonstrate a NOAEL, a LOAEL (lowest-observed-adverse-effect level) may be used. A LOAEL is defined as the lowest experimental dose at which a statistically significant increase in the occurrence of adverse effects is observed.

Five factors may contribute to the determination of the uncertainty and modifying factors. These are (1) the need to accommodate human response variability, including sensitive subgroups; (2) the need to extrapolate from animal exposure data to humans, when human exposure data are unavailable or inadequate; (3) the need to extrapolate from subchronic to chronic exposure data, when the latter are unavailable; (4) the need to account for using a LOAEL, when a NOAEL is unavailable; and (5) the need to extrapolate from a data base that is inadequate or incomplete. Factors between 1- and 10-fold are often used to account for each of these sources of uncertainty.

EPA has adapted the oral RfD methodology to estimate inhalation reference concentrations (RfCs) to be consistent in setting levels of noncarcinogenic chemicals (EPA, 1990). The inhalation RfC methodology departs from the oral RfD paradigm by incorporating dosimetric adjustments to scale the exposure concentration for animals to a human equivalent concentration.

For noncarcinogenic effects, the RfC approach should be the primary method used for setting SMACs. Uncertainty (safety) factors between 1 and 10 should be used for each source of uncertainty listed above, depending on the nature and severity of the adverse effects. For setting SMACs, the duration extrapolation, if any, would be in the opposite direction from that indicated in the third source above, that is, it would be from long-term to short-term exposure. If an exposure-extrapolation approach such as Haber's rule is used and there is considerable uncertainty as to its validity, then the exposure extrapolation should be accompanied by the highest safety factor (10). In addition to the RfC method, alternative methods such as the benchmark dose approach (Crump, 1984a; Chen and Kodell, 1989) should be considered as secondary approaches to setting SMACs.

Carcinogenic Effects

Mathematical Modeling

For carcinogenic effects, particularly those considered to be due to genotoxic events such as alkylation of DNA, a threshold dose may not exist. Beginning with the pioneering work of Mantel and Bryan (1961), attempts have been made to estimate carcinogenic risks on a precise quantitative basis. This task has involved fitting mathematical models to experimental data and extrapolating from these models to predict risks at doses well below the experimental range.

The probit-log dose model initially used by Mantel and Bryan (1961) was a carryover from the tolerance distribution models used for analyzing dose-response mortality data from acute toxicity studies. This model was later replaced by mathematical models based on presumed mechanisms of carcinogenesis. The mechanistic model used most frequently for low-dose extrapolation is the multistage model of Armitage and Doll (1961). A historical perspective on the evolution of the multistage model has been provided by Whittemore and Keller (1978). According to the multistage theory, a malignant cancer cell develops from a single stem cell as a result of a number of biological events (e.g., mutations) that must occur in a specific order. This model predicts that the age-specific cancer incidence rate should increase in proportion to age raised to a power related to the number of stages in the model. This model provides a good description of many forms of human cancer that allow for two to six stages in the carcinogenic process.

Assuming that the rates of transition between stages in the multistage model are linearly related to dose, the dose-response curve for the multistage model is linear at low doses. Low-dose linearity is generally assumed for chemical carcinogens that act through direct interaction with genetic material. It is supported by theoretical considerations in carcinogenesis (Krewski et al., 1989a) and by observations of the linearity of DNA binding with a number of chemical carcinogens at low doses (Lutz, 1990). When carcinogenesis occurs after toxic tissue injury, low-dose linearity may not be applicable, but the assumption is widely used in applications of low-dose risk assessment for regulations in the absence of clear information to the contrary (OSTP, 1985).

EPA (1986) uses the linearized multistage model for low-dose extrapolation. With this procedure, an upper confidence limit, q_1^*, on the coefficient of the linear term in the model is obtained (Crump, 1984b). The value q_1^* represents an upper bound on the slope of the dose-response curve in the low-dose region and on the excess risk above background associated with a unit measure of dose (such as 1 mg/kg of body weight per day). Only an upper confidence limit on the low-dose slope is used by EPA for extrapolation, in part because point estimates of this parameter are highly unstable. Linear extrapolation from an upper confidence limit on excess risk obtained in the experimental range is generally regarded as conservative in the sense of protecting human health. It will be conservative as long as the dose-response relationship in the low-dose region is convex, regardless of its actual mathematical formulation (Gaylor and Kodell, 1980).

Recently, a two-stage model that incorporates tissue growth and cellular kinetics has been proposed as an alternative to the multistage model for cancer risk assessment (Moolgavkar and Venzon, 1979; Moolgavkar and Knudson, 1981). Mathematical formulations of this stochastic birth-death-mutation model have been given by Moolgavkar and Luebeck (1990) and by Greenfield et al. (1984). The model assumes that two genetic events (mutations), each occurring at the time of cell division, are necessary for a normal cell to become malignant. Allowance is made for natural growth of the target tissue and for clonal expansion of the pool of cells that have undergone the first mutation.

The birth-death-mutation model provides a convenient framework for describing initiation-promotion-progression mechanisms of carcinogenesis. Initiating activity may be quantified in terms of the rate of occurrence of the first mutation. The rate of occurrence of the second mutation describes progression to a fully differentiated cancerous lesion. Promotional activity is measured by the difference in the birth and death rates of initiated cells. Carcinogenic agents may be classified as initiators, promoters, or progressors according to components of the model that they are presumed to affect (Thorslund et al., 1987).

Application of the two-stage model with clonal expansion in risk assessment for putative tumor-promoting agents has been investigated on a limited basis (Thorslund and Charnley, 1988). Before this model can be recommended for routine application, its statistical properties require further study, particularly with respect to predictions of risk at low doses (Portier, 1987). In the absence of promotional effects and variability in the pool of normal cells, the two-stage birth-death-mutation model is reduced to the two-stage version of the classical multistage model.

Pharmacokinetic Considerations

Chemical carcinogens may require some form of metabolic activation to exert their effects. If metabolic activation can be characterized adequately in terms of a pharmacokinetic model, then the dose delivered to the target should be used in place of the administered dose for purposes of dose-response modeling and low-dose extrapolation (Hoel et al., 1983). In general, the use of delivered rather than administered doses may be expected to lead to more accurate predictions of carcinogenic risk (NRC, 1987b). Such predictions, however, may introduce substantial uncertainty if complicated, physiologically based pharmacokinetic models with 20 to 40 parameters, each subject to uncertainty, are used for tissue dosimetry (Portier and Kaplan, 1989; Farrar et al., 1989).

Intermittent Exposure

Most bioassay data suitable for low-dose extrapolation reflect continuous exposure to a carcinogen, necessitating the translation of risks calculated for continuous exposures to risks associated with short-term or intermittent exposures. Crump and Howe (1984) developed a methodology for applying the multistage model to carcinogenic risk assessment when exposure to a carcinogen is intermittent. Their results have been adapted by the Committee on Toxicology (COT) (NRC, 1986b) for setting emergency and short-term exposure guidelines for chemicals of interest to the U.S. Department of Defense. Both multistage and two-stage models have been studied with respect to age at which exposure occurs (Kodell et al., 1987; Chen et al., 1988; Murdoch and Krewski, 1988; Krewski and Murdoch, 1990). The results of these investigations indicate that when an early stage in the carcinogenic process is dose-dependent, early exposures will be of greater concern than later exposures. Conversely, when a late stage is dose-dependent, late-life exposures pose a higher risk. In the context of the classical multistage model, the potential increased excess risk from exposure for a fraction, f, of a lifetime at a given dose rate, d, will never be more than k times the excess risk from full lifetime exposure at dose rate fd, where k is the number of stages in the model. (Note that administration of daily dose d for fraction f of a lifetime yields the same cumulative lifetime dose as administration of dose fd over the course of a lifetime.) With the two-stage model, however, the increased risk can be substantial when the agent of interest greatly increases the proliferation rate of the initiated cell population.

Carcinogenic Mixtures

The interactive effects between two carcinogens have been extensively studied (Krewski and Thomas, 1992). However, the existence of synergism at low levels of exposure cannot be assessed directly. The multistage and the multiplicative relative-risk models for cancer predict near additivity of excess risk at low doses, even when there is nonadditivity at higher doses (NRC, 1988c; Krewski et al., 1989b; Chen et al., 1990). Kodell et al. (1991) recently described departures from additivity within the context of the two-stage birth-death-initiation model. Additivity is expected, however, at low doses for initiators and progressors, as well as nongenotoxic promoters that may demonstrate a threshold.

In the absence of data from experiments involving joint exposure to multiple agents, the health risk from joint exposure to two or more carcinogens must be based on the results of experiments with single compounds. In the absence of evidence to the contrary, the total risk from a mixture of carcinogens generally is assumed to equal the sum of the individual risks of the components (EPA, 1986). This assumption is consistent with the concept of a group-limit value, which implies that individual allowable concentrations must be reduced when they are in a mixture of chemical toxicants with the same or similar effects to control the overall risk. Chen et al. (1990) noted that if upper confidence limits on individual estimates of carcinogenic risk of the same order of magnitude simply are summed as an approximation to an upper confidence limit on the total excess risk of a mixture, the upper bound will be conservative. Using the assumption of additivity of component risks to overcome this conservatism, Chen et al. (1990) developed a formal procedure for calculating an upper confidence limit on the total carcinogenic risk of a mixture.

Interspecies Extrapolation

Quantitative methods for cancer risk assessment for carcinogenicity must also take into account species differences. Traditionally, dose equivalency among species has been based on body weight. However, many physiological constants (e.g., consumption of water, food, and oxygen) vary as a power function of body weight (NRC, 1987c). Dourson and Stara (1983) pointed out that toxicity of many compounds may be related more closely to body surface area than to body weight (Pinkel, 1958; Freireich et al., 1966). The use of a general power function of body weight for species conversion includes body-weight conversion as a special case with a power of unity. Because body surface area varies approximately in proportion to the 2/3 power of body weight, surface-area conversion is another special case of the general power-law relation. Based on a reanalysis of the data of Freireich et al. (1966), Travis and White (1988) suggested that body weight to the 3/4 rather than to the 2/3 power may be more appropriate for species conversion. Allen et al. (1988) found a high correlation of carcinogenic potency in animals and humans but were unable to resolve whether body weight or surface area provided a better basis for species conversion.

If the compound is well defined in its metabolism and pharmacokinetics, these data should be used in preference to the body-weight to surface-area correction. The body-weight to surface-area correction is essentially a correction for metabolic rate based on body size. The basis of this correction is more related to rates of basal metabolism than of xenobiotic metabolism. Thus, in the face of more specific data on the comparative pharmacokinetics and metabolism of the chemical in question, the correction should not be used.

Extrapolation of Animal Data to Humans

Interpretation of data derived from animal experiments requires experienced scientific judgment in a variety of disciplines. Consideration should be given to the conditions under which the data were obtained and their relevance to conditions of human exposure under question. How similar are the test species and the test organ to humans and the corresponding human organ in morphology, in sensitivity of response to the contaminant, and in metabolism and disposition of the contaminant? Were the observed responses of animals the consequences of exposure conditions to which humans may be subjected?

Development of SMACs requires that animal data be extrapolated to humans. The species most representative of humans, considering both toxicological and pharmacokinetic characteristics, should be used for determining the appropriate exposure limit. If data are not available on which species best represents humans, it is prudent to use data from the most sensitive animal model to set appropriate limits.

As discussed earlier, uncertainty factors can be used when an estimate of the NOAEL in animals is available. In the application of uncertainty factors, particularly susceptible individuals should be considered. People at unusual risk might include those at risk because of unusual physical exertion and other stresses; such risk might occur to astronauts. Persons of suspected high susceptibility (i.e., very old or those with known diseases) are unlikely to be space-station workers, thus reducing the need for safety factors to allow for these states or conditions. The value of the uncertainty factor also will depend on the quality of the animal data used to determine the NOAEL. The RfD often has been set at a level 100-fold below the NOAEL derived from lifetime animal experiments. Larger uncertainty factors also have been used; for example, a 1,000-fold factor usually is used to establish an RfD when only subchronic (90-day) toxicity studies are available. Higher uncertainty factors also may be used because of the altered physiological status of astronauts in the space shuttle or space station.

Conversion from animals to humans may be done on a body-weight or surface-area basis, as discussed earlier. When available, pharmacokinetic data on tissue doses also may be considered for use in species conversion. In using toxicological data, data may have to be extrapolated from those on oral exposure of animals (in milligrams per kilogram of body weight) and converted to inhalation exposure of humans (in milligrams per cubic meter of air). Two approaches are possible: extrapolation from rat oral data to human oral dose and then to human inhalation concentration; or extrapolation from rat oral data to rat inhalation concentration and then to human inhalation concentration. The first approach does not assume similar breathing rates for animals and humans. A 70-kg man breathes air at 7-10 L/min at rest, 20 L/min when moderately active, and 40-60 L/min when engaged in heavy work. A resting 250-g (0.250-kg) rat breathes air at 200 ml/min (0.2 L/min).

To illustrate the first approach, assume that an oral dose of 1 mg/kg of body weight per day in rats is equivalent to an oral dose of 1 mg/kg of body weight per day in humans. A 70-kg man at rest inhales 15 m^3 of air over a 24-hr period. The concentration inhaled by a 70-kg man in 1 day is shown by the equation [(1 mg/kg)(70 kg)]/(15 m^3) = 4.7 mg/m^3. However, not all of what is inhaled will be absorbed. Toxicokinetic information available on the uptake of the compound of interest or on similar compounds should be used to correct the above equation. In the above example, if only 50% of an inhaled concentration of a material is deposited or absorbed by humans, then an individual would have to be exposed to a concentration of 9.4 mg/m^3 over a 24-hr period to achieve the equivalent of an oral dose of 1 mg/kg per day.

To illustrate the second approach, suppose that a 250-g rat at rest inhales air at 200 ml/min or 0.288 m^3 over a 24-hr period. An oral dose of 1 mg/kg in the rat is equivalent to an inhalation concentration in the rat, as shown by [(1 mg/kg)(0.250 kg)]/(0.288 m^3) = 0.87 mg/m^3. For a 70-kg man inhaling 15 m^3 of air in 24 hr, the conversion to an inhalation concentration is (0.87 mg/m^3)(0.288 m^3/0.250 kg)(70 kg/15 m^3) = 4.7 mg/m^3. Again, corrections should be made for the percentage of inhaled material deposited or absorbed.

SMACs for Carcinogens

The determination of SMACs for carcinogens generally will require extrapolation of toxicological or epidemiological data obtained under conditions of long-term (often lifetime) exposure to periods of shorter duration, such as 1, 30, or 180 days. Here, we illustrate how the methods proposed by Kodell et al. (1987) based on the multistage model can be used to set SMACs for carcinogens.

Suppose that data from a long-term animal bioassay or human epidemiological study have been used to arrive at an average daily lifetime exposure level, d, of a particular chemical, corresponding to a particular lifetime excess carcinogenic risk. As reported by COT, a lifetime risk level of 10^{-4} has been used by the U.S. Department of Defense (NRC, 1986b) to set exposure guidelines for carcinogenic chemicals. Assume that d has been estimated with the use of any available pharmacokinetic information to determine tissue dose, the linearized multistage model, and an appropriate interspecies conversion factor.

Suppose that the carcinogen of interest affects only a single stage of a multistage process and that there are up to $k = 6$ stages in total. Let i $(i = 1, 2, ..., k)$ denote the index of the stage that is dose-related. Consider the case of an astronaut exposed to a constant daily dose, D, from age t_0 to t_1. Using results of Kodell et al. (1987, p. 340), it can be shown that exposure to a dose,

$$D = \frac{dt^k}{\sum_{j=1}^{i} \binom{k}{j-1} \left[t_0^{j-1}(t - t_0)^{k-j+1} - t_1^{j-1}(t - t_1)^{k-j+1} \right]}, \quad (1)$$

from time t_0 to time t_1 ($0 \leq t_0 < t_1 \leq t$) will yield the same excess risk as a constant daily dose, d, from birth to age t, where t denotes the time when excess risk is evaluated (usually the average human lifetime).

To illustrate, consider a three-stage process with only the first stage dose-related, and suppose that exposure occurs at the earliest age possible. The assumed earliest age of exposure for astronauts will be age 25 (9,125 days), and the assumed average human lifetime will be 70 years (25,550 days). To determine the 180-day SMAC, given the acceptable lifetime daily dose d, the value of D in Eq. 1 is evaluated with $k = 3$, $t = 25,550$, $t_0 = 9,125$, and $t_1 = 9,125 + 180 = 9,305$. This gives $D = 115.8\ d$. For a 30-day SMAC, $k = 3$, $t = 25,550$, $t_0 = 9,125$, and $t_1 = 9,155$. Substituting these values into Eq. 1 gives $D = 688.2\ d$.

An alternative but equivalent approach is to calculate an intermediate dose, D^*, that would yield the same total exposure as the average daily lifetime exposure level d, and then apply an adjustment factor, f, based on the multistage model. This approach has been followed by COT (NRC, 1986b) in performing risk assessments for the U.S. Department of Defense. With this approach,

$$D^* = dt/(t_1 - t_0), \quad (2)$$

which is simply the average daily dose from age t_0 to t_1 that would yield the same cumulative dose as continuous exposure for a lifetime to a daily dose of d. The SMAC is then defined by D^*/f, where

$$f = \frac{\{t/(t_1-t_0)\}\left\{\sum_{j=1}^{i}\binom{k}{j-1}\left[t_0^{j-1}(t-t_0)^{k-j+1} - t_1^{j-1}(t-t_1)^{k-j+1}\right]\right\}}{t^k}.$$

(3)

In the preceding example, D^* would be 141.9 d for the 180-day SMAC and 851.7 d for the 30-day SMAC. The value of f would be 1.23 for the 180-day SMAC and 1.24 for the 30-day SMAC. Considering both the multistage and two-stage models, Murdoch et al. (in press) suggest that the value of f for astronauts between 25 and 45 years of age is unlikely to exceed a value of ≈ 2.

The case of near instantaneous exposure (e.g., single-day exposure) at time t_0 to a total dose D requires separate treatment. In this limiting case, SMACs for exposure durations of 1 day or less may be based on the relationship

$$D = \frac{dt^k}{\left\{\sum_{j=1}^{i}\binom{k}{j-1}\left[(k-j+1)t_0^{j-1}(t-t_0)^{k-j} - (j-1)t_0^{j-2}(t-t_0)^{k-j+1}\right]\right\}},$$

(4)

which applies in the case of a carcinogen that affects only the ith stage of a k-stage process. The corresponding adjustment factor for the time-weighted average dose, $D^* = dt$, is

$$f = \frac{\left\{\sum_{j=1}^{i}\binom{k}{j-1}\left[(k-j+1)t_0^{j-1}(t-t_0)^{k-j} - (j-1)t_0^{j-2}(t-t_0)^{k-j+1}\right]\right\}}{t^{k-1}}.$$

(5)

It must be remembered that extrapolation from a daily lifetime exposure level and conversion to an instantaneous exposure level using Eq. 4 or Eq. 5 is an extreme case and is valid only under the assumptions underlying the multistage theory of carcinogenesis. As in Eq. 3, the value of f based on Eq. 5 is unlikely to exceed 2 (Murdoch et al., in press).

Use of OSHA, ACGIH, NRC, or Other Limits

There are several inhalation guidelines that can be used as sources of information on compounds of interest. The guidelines, established by OSHA,[a] ACGIH,[b] EPA, ATSDR,[c] and NRC, should be reviewed before establishing SMACs.

The subcommittee has reviewed reports of analyses of atmospheric contaminants detected in a variety of closed environments: seven manned spaceflights (Mercury and Gemini series), many space-shuttle flights, several ground-based simulated cabin atmospheres (both manned and unmanned), nuclear submarine and Sealab experience, and analyses of off-gas products from cabin materials. Nearly 300 compounds that could be possible space-station contaminants have been identified.

Many of these compounds have been reviewed by the NRC's COT, and 1- and 24-hr emergency exposure guidance levels (EEGLs) or 90-day continuous exposure guidance levels (CEGLs) for submarines have been recommended. The recommended EEGLs are similar to 1- and 24-hr SMACs and permit some reversible effects, such as headache and mild irritation. The differences between the conditions for purging the atmospheres of spacecraft and of submarines must be considered. Although designing for a complete atmospheric purge by opening the spacecraft may be possible, the need for such an emergency operation must be minimized. In contrast, submarines can surface to ventilate except in extraordinary circumstances.

EEGLs differ from STELs (short-term exposure limits) recommended by OSHA or ACGIH in that STELs are generally 15-min limits to which workers may be exposed daily for many years. The 1- and 24-hr SMACs, on the other hand, are limited for infrequent emergency exposure.

The continuous exposure guidance levels (CEGLs) recommended by COT for submarines are ceiling concentrations designed to avoid adverse health effects, either immediate or delayed, of more prolonged exposures and to avoid degradation in crew performance that might endanger the objectives of a particular mission as a consequence of continuous exposure for up to 90 days.

---

[a] Occupational Safety and Health Administration.
[b] American Conference of Governmental Industrial Hygienists.
[c] Agency for Toxic Substances and Disease Registry.

ATSDR is developing documents on several industrial chemicals. These documents provide detailed background information on the chemicals and the basis for setting limits for those chemicals.

EPA has developed numerous cancer assessments and inhalation RfCs that are accessible on EPA's Integrated Risk Information System (IRIS). Both values assume a 70-year exposure, which would require adjustment to the space scenario. EPA's cancer assessment methodology has been described in *The Risk Assessment Guidelines of 1986* (EPA, 1987).

All the documents used to establish previous industrial or public exposure limits for airborne materials should be reviewed for pertinent information before establishing SMACs. The purpose of this comparison is not to mimic the regulatory levels set by others, but to determine if the SMACs are reasonable in light of the special needs of NASA. The background information on how other agencies set guidance levels also should be useful in setting SMACs.

Exposure to Mixtures

Individual spacecraft limits for single compounds have been established for guidance in engineering design of atmosphere-handling systems and represent an upper limit for each compound without regard to its potential occurrence in mixtures or in the presence of other toxicants. However, atmospheric contaminants are encountered most often as complex mixtures, and the toxicological hazard to humans from inhalation, especially on a long-term basis, must be assessed in terms of permissible atmospheric loadings for these mixtures. Therefore, individual limits must be integrated into sets of group limits to reflect air-quality conditions judged to be safe for humans during given exposure periods.

The spacecraft atmosphere consists of a mixture of compounds, many of which have similar effects that are likely to be additive. These potential additive effects must be considered in the assessment of the toxicological hazard of contaminant mixtures. The following guidelines for SMAC development provide for the potential summation of toxic effects of contaminants.

1. The concentration of each contaminant in the spacecraft atmosphere must not exceed its SMAC (1 hr, 24 hr, or 180 days) value

expressed as a time-weighted average or as a ceiling value, as appropriate.

2. For each one of the groups of contaminants, a group-limit concept will be utilized to evaluate the toxicological hazard of the group. In each group, the summation, T, of the ratios of concentration to the SMAC value of each member of the group must not exceed unity (ACGIH, 1989). The following formula will apply when the components in a mixture from structurally related compounds have similar toxicological effects:

$$\frac{C_1}{SMAC_1} + \frac{C_2}{SMAC_2} + \frac{C_3}{SMAC_3} + \ldots + \frac{C_n}{SMAC_n} = T \leq 1, \quad (6)$$

where C_i is the observed concentration of the ith material and $SMAC_i$ is the corresponding permissible concentration.

Example: Air contains 400 ppm of acetone (SMAC, 750 ppm), 150 ppm of sec-butyl acetate (SMAC, 200 ppm), and 100 ppm of methyl ethyl ketone (SMAC, 200 ppm).

$$\frac{400}{750} + \frac{150}{200} + \frac{100}{200} = 0.53 + 0.75 + 0.5 = 1.78. \quad (7)$$

Although each individual exposure does not exceed its SMAC, the permissible total is exceeded, and exposures must be reduced (ACGIH, 1989).

There will be situations in which the effects of exposure to mixtures will be greater than additive, and this information must be taken into account in setting SMACs.

Physiological adaptive responses that are important in space operation may involve muscle remodeling, loss of red-blood-cell mass (anemia), altered nutritional requirements, and behavioral changes. The panoply of all physiological changes considered important demonstrates that the individual astronaut is in an altered homeostatic state. How this altered state modifies reactions to chemicals in the space-station environment requires additional information. With current knowledge, the maximum allowable concentrations of airborne chemicals in the closed system of space travel would have to be reduced from permissible levels on earth. The astronaut in space is an atypical human, particularly following extended residence in space.

Some spacecraft contaminants may present a carcinogenic risk. In this case, SMACs should satisfy the following guidelines:

1. The lifetime excess risk of cancer resulting from exposure to a chemical contaminant in space for periods up to 180 days should not exceed 10^{-4} or other levels acceptable to NASA. The lifetime risk associated with such short-term exposures may be estimated on the basis of a time-weighted average lifetime dose in Eq. 2, divided by an adjustment factor of $f = 2$.

2. The lifetime risk for joint exposure to two or more carcinogens should not exceed 10^{-4} or other levels acceptable to NASA. Assuming additivity of risks, the total risk is determined by summing the individual risks.

Modification of Contaminant Toxicity by Environmental Factors

The special conditions of the space environment must be taken into account in defining spacecraft contaminant standards. Examples of potential interaction between contaminants and environmental factors include the effects of increased levels of radiation on the radiomimetic action of benzene and the effects of increased fatigue from stress factors (e.g., noise, vibration, and weightlessness (Kaplan, 1979)). Lung function may be different, deposition of particles is different under microgravity conditions, the potential toxic effects of inhaled particles on CNS active agents may be different, and the toxicity of gases may be altered by condensation on particulates in the spacecraft.

Astronauts will be physically, physiologically, and psychologically compromised for the following reasons: loss of muscle mass, loss of bone mass, depressed immune system, decreased red-blood-cell mass (anemia), altered nutritional requirements, behavioral changes from stress, fluid shift in the body, altered hormonal status, and altered drug metabolism.

These changes imply that an astronaut in space will be in an altered homeostatic state and may be more susceptible to toxic chemicals.

The adaptive physiological changes that occur in microgravity may prove to be detrimental to astronauts. For example, Merrill et al. (1990) analyzed a number of liver enzymes from rats that had flown aboard Cosmos 1887. They found a decrease in amounts of cytochrome P-450. In addition, decreases occurred in the enzyme aniline hydroxylase and ethylmorphine N-demethylase. Since this family of enzymes is responsible for the metabolism of steroid hormones and a variety of xenobiotics, including antibiotics and drugs, this observation could have important implications for space medicine and toxicology.

The changes in drug-metabolizing enzymes, immunological changes, and other alterations could be important factors in disease susceptibility and chemical toxicity, and those should be taken into account to recommend SMACs.

Risk assessment of chemicals is based on biological studies conducted in animals and humans on earth. It is unclear how much of the data can be generalized to the space environment. A variety of assumptions will be necessary. The quality of these assumptions can be improved substantially as more precise human data are gained from astronauts. Therefore, it is important to take into account the physiological changes induced in the space crew and the impact of these changes on SMAC values for various contaminants.

GENERAL APPROACH TO ESTABLISHING SMACs

The first step in producing a document describing the SMAC for a chemical is to collect and review all relevant information available on the compound. Based on the review of the literature, SMACs for different time periods are developed and a rationale is provided for each recommendation. The next step is to review and evaluate prior EEGLs and CEGLs, ACGIH's threshold limit values (TLVs), OSHA's recommendation for permissible exposure limits (PELs), NIOSH's (National Institute of Safety and Health) recommended exposure limits (RELs), ATSDR's recommendations, and EPA's ambient air quality standards. The intent of such a review is not to mimic the regulatory levels set by others but to make use of all available background information gathered by other agencies in evaluating the health effects of the chemical of interest.

Development of 1- and 24-hr SMACs for different durations of exposure usually begins with providing a SMAC for the shortest exposure of 1 hr. Values for 24-hr SMACs typically are developed by using Haber's law when applicable. Important determinants of the nature and severity of biological effects resulting from exposure to a chemical are the level and duration of exposure. There has been a general interest in assessing the extent to which the concentration, C, of a given chemical and the duration of exposure, T, interrelate to determine the magnitude of the biological response, K. The overall relationship between C, T, and K is a complex, three-dimensional surface. However, if interest is in a particular level of biological response, for example, $K = k$, the problem becomes two-dimensional.

By restricting the biological response to a particular level, k, Haber postulated that the formula $C \times T = k$ could be used to relate the toxic effect of certain inhalable substances to their concentration and time of exposure. Although the above formula, which sometimes is referred to as Haber's law, is clearly valid only for a limited number of substances and only for certain combinations of concentration and exposure time, it has been used to synthesize data relating concentration and time of exposure. A more general expression for examining relationships between C, T, and k is given by $C^a \times T^b = k$, where the exponents a and b are estimated from the data. This formula allows for the fact that C and T do not always contribute equally to the observed toxicity.

When detoxification or recovery occurs and data are available on 24-hr exposures, these factors must be taken into account in modifying Haber's law ($C \times T = k$). Haber's law is inappropriate for materials, such as ammonia and NO_2, that have been shown to be more toxic with high concentrations over short periods. High concentrations of particulates also have been shown to overload normal host defense systems, making predictions from such data bases unreliable for long-term, low-level exposures. Consequently, caution must be exercised in using even the general expression for C, T, and k when evaluating exposure conditions that are likely to result in comparable effects in laboratory animals and humans.

The approach to establishing 1- or 24-hr SMACs is judged on a substance-by-substance basis. For substances that affect several organ systems or have multiple effects, all end points—including reproductive (in both sexes), developmental, carcinogenic, neurotoxic, respiratory, and other organ-related effects—are evaluated, the most important or most sensitive effects receiving the major attention. For the many compounds, such as upper respiratory tract irritants, for which there are reliable human data and a high degree of confidence that effects from a single exposure will be reversible, then NASA may establish exposure limits directly from the available data. For those compounds that have few human data but sound animal data and effects that are likely to be reversible following a single exposure, a species extrapolation factor may be employed, the magnitude of which will depend on the quality of the data.

In the absence of better information, a safety factor of 10 is suggested for 1- and 24-hr SMACs when only animal data are available and extrapolation from animals to humans is necessary for acute, short-term effects or when the likely route of human exposure differs from that of a relevant experiment. With carcinogenic chemicals, the SMAC is set so that the estimated risk of a neoplasm in the upper 95% confidence limit is no more than 1 in 10^4 or some other risk acceptable to NASA.

SMACs do not represent a sharp dividing line between safe and unsafe concentrations. Some people are expected to be affected adversely when SMACs are exceeded. However, some persons may be affected adversely even when SMACs are not exceeded. When exposure lasts longer than 24 hr, the 24-hr SMAC no longer applies, and appropriate measures must be taken to comply with the concentration implied by the corresponding 7-day, 30-day, and 180-day SMAC values.

The 30- and 180-day SMACs are generally 0.01-0.1 times the 24-hr SMAC for noncarcinogenic substances. When detoxification is substantial, a division by 10 may be more appropriate. When detoxification is not evident or is slow, a division by 100 or more may be more appropriate. When the substance accumulates in tissues, as do halogenated biphenyls and metals, even larger safety factors may be used. In other cases, lower safety factors may be applied, depending on the data. The choice within these general guidelines must be determined for each material separately. When data from chronic studies are available, they are used to derive 180-day SMACs, applying safety factors as needed. With carcinogenic chemicals, an estimate of risk is provided for the recommended SMACs. When a substance is known to cause an effect that will be aggravated by microgravity, additional safety factors should be used.

The suggested format for a SMAC document is shown in Appendix 1. The conversion factors and the reference values used in establishing SMACs are listed in Appendixes 2 and 3, respectively.

References

ACGIH (American Conference of Governmental Industrial Hygienists). 1989. P. 18 in Documentation of the Threshold Limit Values and Biological Exposure Indices, 6th Ed. Cincinnati, Ohio: American Conference of Governmental Industrial Hygienists.

Adams, J.M., and S. Cory. 1991. Transgenic models of tumor development. Science 254:1161-1167.

Albertini, R.J., J.A. Nicklas, J.P. O'Neill, and S.H. Robison. 1990. In vivo somatic mutations in humans: Measurement and analysis. Annu. Rev. Genet. 24:305-326.

Aldridge, W.N. 1986. The biological basis and measurement of thresholds. Annu. Rev. Pharmacol. Toxicol. 26:39-58.

Allen, B.C., K.S. Crump, and A.M. Shipp. 1988. Correlation between carcinogenic potency of chemicals in animals and humans. Risk Anal. 8:531-544.

Anger, W.K. 1984. Neurobehavioral testing of chemicals: Impact on recommended standards. Neurobehav. Toxicol. Teratol. 6:147-153.

Armitage, P., and R. Doll. 1961. Stochastic models for carcinogenesis. Pp. 19-38 in Proceedings of the Fourth Berkeley Symposium on Mathematical Statistics and Probability, Vol. 4, J. Neyman, ed. Berkeley, Calif.: University of California Press.

Barnes, D.G., and M.L. Dourson. 1988. Reference dose (RfD): Description and use in health risk assessments. Regul. Toxicol. Pharmacol. 8:471-486.

Borgaonkar, D.S. 1991. Chromosomal Variation in Man. A Catalog of Chromosomal Variants and Anomalies, 6th Ed. New York: Wiley-Liss.

Brandriff, B.F., and L.A. Gordon. 1990. Human sperm cytogenetics and the one-cell zygote. Pp. 183-192 in Biology of Mammalian Germ Cell Mutagenesis. Banbury Report 34, J.W. Allen, B.A. Bridges, M.F. Lyon, M.J. Moses, and L.B. Russell, eds. Cold Spring Harbor, N.Y.: Cold Spring Harbor Laboratory Press.

Calloway, D.H. 1968. Gas in the alimentary canal. Pp. 2839–2859 in Handbook of Physiology, Section 6, Vol. 5. Alimentary Canal, C.F. Code, ed. Washington, D.C.: American Physiological Society.

Calloway, D.H. 1971. End products of human metabolism as affected by diet and space conditions. Environ. Biol. Med. 1:197–202.

Calloway, D.H., and E.L. Murphy. 1969. Intestinal hydrogen and methane of men fed space diet. Pp. 102–109 in Proceedings of the Open Meeting of Working Group V at the Eleventh Plenary Meeting of COSPAR, Tokyo, May 10, 1968, Life Sciences and Space Research, Vol. 7, W. Vishniac and F.G. Favorite, eds. Amsterdam: North-Holland.

Checkoway, H., N. Pearch, and D.J. Crawford-Brown. 1989. Research Methods in Occupational Epidemiology. New York: Oxford University Press. 344 pp.

Chen, J.J., and R.L. Kodell. 1989. Quantitative risk assessment for teratological effects. J. Am. Stat. Assoc. 84:966–971.

Chen, J.J., R.L. Kodell, and D.W. Gaylor. 1988. Using the biological two-stage model to assess risk from short-term exposures. Risk Anal. 8:223–230.

Chen, J.J., D.W. Gaylor, and R.L. Kodell. 1990. Estimation of the joint risk for multiple-compound exposure based on single-compound experiments. Risk Anal. 10:285–290.

Cogoli, A., M. Valluchi-Morf, M. Mueller, and W. Briegleb. 1980. Effect of hypogravity on human lymphocyte activation. Aviat. Space Environ. Med. 51:29–34.

Crump, K.S. 1984a. A new method for determining allowable daily intakes. Fundam. Appl. Toxicol. 4:854–871.

Crump, K.S. 1984b. An improved procedure for low-dose carcinogenic risk assessment from animal data. J. Environ. Pathol. Toxicol. Oncol. 5:339–348.

Crump, K.S. and R.B. Howe. 1984. The multistage model with a time-dependent dose pattern: Applications to carcinogenic risk assessment. Risk Anal. 4:163–176.

Deitrick, J.E., G.D. Whedon, E. Shorr, V. Toscani, and V.B. Davis. 1948. Effects of immobilization upon various metabolic and physiologic functions of normal men. Am. J. Med. 4:3–36.

Dews, P.B. 1953. The measurement of the influence of drugs on voluntary motor activity in mice. Br. J. Pharmacol. 8:46–48.

Dourson, M.L. 1986. New approaches in the derivation of acceptable daily intake (ADI). Comments Toxicol. 1:35–48.

Dourson, M.L. and J.F. Stara. 1983. Regulatory history and experimental support of uncertainty (safety) factors. Regul. Toxicol. Pharmacol. 3:224–238.

Durnova, G.N., A.S. Kaplanski, and V.V. Portugalov. 1977. Effect of a 22-day space flight on the lymphoid organs of rats [in Russian]. Kosm. Biol. Aviakosm. Med. 11:53–57.

EPA (U.S. Environmental Protection Agency). 1985. Toxic Substances Control Act test guidelines: Final rules. Fed. Regist. 50(188):39426–39428; 39433–39434.

EPA (U.S. Environmental Protection Agency). 1986. Guidelines for carcinogen risk assessment. Fed. Regist. 51(185):33992–34003.

EPA (U.S. Environmental Protection Agency). 1987. The Risk Assessment Guidelines of 1986. EPA 600/8-87/045. Washington, D.C.: U.S. Environmental Protection Agency.

EPA (U.S. Environmental Protection Agency). 1988a. Proposed guidelines for assessing male reproductive risk. Fed. Regist. 53(126):24850–24869.

EPA (U.S. Environmental Protection Agency). 1988b. Proposed guidelines for assessing female reproductive risk. Fed. Regist. 53(126):24834–24847.

EPA (U.S. Environmental Protection Agency). 1990. Interim Methods for Development of Inhalation Reference Concentrations. EPA 600/8-90/066A. Environmental Criteria and Assessment Office. Research Triangle Park, N.C.: U.S. Environmental Protection Agency.

EPA (U.S. Environmental Protection Agency). 1991. Guidelines for developmental toxicity risk assessment. Fed. Regist. 56(234):63798–63826.

FAA (Federal Aviation Administration). 1988. System Design and Analysis. Advisory Circular 25.1309-1A. Washington, D.C.: U.S. Department of Transportation.

Farrar, D., B. Allen, K. Crump, and A. Shipp. 1989. Evaluation of uncertainty in input parameters to pharmacokinetic models and the resulting uncertainty in output. Toxicol. Lett. 49:371–385.

Fenters, J.D., J.C. Findlay, C.D. Port, R. Ehrlich, and D.L. Coffin. 1973. Chronic exposure to nitrogen dioxide: Immunologic, physiologic, and pathologic effects in virus-challenged squirrel monkeys. Arch. Environ. Health 27:85–89.

Freireich, E.J., E.A. Gehan, D.P. Rall, L.H. Schmidt, and H.E. Skipper. 1966. Quantitative comparison of anticancer agents in mouse, rat, hamster, dog, monkey and man. Cancer Chemother. Rep. 50:219-244.

Gaylor, D.W., and R.L. Kodell. 1980. Linear interpolation algorithm for low dose risk assessment of toxic substances. J. Environ. Pathol. Toxicol. 4(5&6):305-312.

Gmunder, F.K., G. Lorenzi, B. Bechler, P. Joller, J. Muller, W.H. Zeigler, and A. Cogoli. 1988. Effect of long-term physical exercise on lymphocyte reactivity: Similarity to spaceflight reactions. Aviat. Space Environ. Med. 59:146-151.

Goldsmith, J.R., G. Patashnik, and R. Israeli. 1984. Reproductive outcomes in families of DBCP-exposed men. Arch. Environ. Health 39:85-89.

Gooch, P.C., and C.A. Berry. 1969. Chromosome analyses of Gemini astronauts. Aerospace Med. 40:610-614.

Greenfield, R.E., L.B. Ellwein, and S.M. Cohen. 1984. A general probabilistic model of carcinogenesis: Analysis of experimental urinary bladder cancer. Carcinogenesis 5:437-445.

Hanley, T.R., Jr., J.T. Young, J.A. John, and K.S. Rao. 1984. Ethylene glycol monomethyl ether (EGME) and propylene glycol monomethyl ether (PGME): Inhalation fertility and teratogenicity studies in rats, mice, and rabbits. Environ. Health Perspect. 57:7-12.

Hoel, D.G., N.L. Kaplan, and M.W. Anderson. 1983. Implication of nonlinear kinetics on risk estimation in carcinogenesis. Science 219:1032-1037.

Holick, M.F. 1986. Vitamin D requirements for the elderly. Clin. Nutr. 5:121-129.

Holick, M.F. 1987. Photosynthesis of vitamin D in the skin: Effect of environmental and lifestyle variables. Fed. Proc. Fed. Am. Soc. Exp. Biol. 46:1876-1882.

Holson, R.R., M.G. Paule, and F.M. Scalzo, eds. 1990. Methods in behavioral toxicology and teratology. Neurotoxicol. Teratol. 12(5).

Humphries, W.R., J.L. Reuter, and R.G. Schunk. 1990. Space Station Freedom Environmental Control and Life Support System Design—A Status Report. SAE Technical Paper Series No. 901211. Warrendale, Pa.: Society of Automotive Engineers.

Johnson, P.C., T.B. Driscoll, and A.D. LeBlanc. 1974. Blood volume changes. Pp. 495-505 in Proceedings of the Skylab Life Sciences Symposium, August 27-29, 1974, Vol. 2. NASA TM X-58154. Houston, Tex.: Lyndon B. Johnson Space Center, National Aeronautics and Space Administration.

Kaplan, H.L. 1979. Contaminants. Pp. 17–56 in The Physiological Basis for Spacecraft Environmental Limits, J.M. Waligora, coordinator. NASA Ref. Publ. 1045. Houston, Tex.: Lyndon B. Johnson Space Center, National Aeronautics and Space Administration.

Kay, H.H., and D.R. Mattison. 1985. How radiation and chemotherapy affect gonadal function. Contemp. Ob. Gyn. 26:109–127.

Kimmel, C.A., and J. Buelke-Sam. 1988. Current status of behavioral teratology science and regulation. CRC Crit. Rev. Toxicol. 19:1–10.

Kimzey, S.L. 1977. Hematology and immunology studies. Pp. 249–282 in Biomedical Results from Skylab, R.S. Johnston and L.F. Dietlein, eds. Washington, D.C.: National Aeronautics and Space Administration.

Klaassen, C.D. 1986. Principles of toxicology. Pp. 11–32 in Casarett and Doull's Toxicology: The Basic Science of Poisons, 3rd Ed., C.D. Klaassen, M.O. Amdur, and J. Doull, eds. New York: Macmillan.

Knight, D.R., J. O'Neill, S.M. Gordon, E.H. Luebcke, and J.S. Bowman. 1984. The Body Burden of Organic Vapors in Artificial Air: Trial Measurements Aboard a Moored Submarine. Memo Report 84-4. Groton, Conn.: U.S. Naval Medical Research Laboratory.

Knight, D.R., J. O'Neill, J.S. Bowman, and S.M. Gordon. 1985. Use of the expired breath to monitor the air quality in sealed capsules [abstract]. Physiologist 28:336.

Kodell, R.L., D.W. Gaylor, and J.J. Chen. 1987. Using average lifetime dose rate for intermittent exposures to carcinogens. Risk Anal. 7:339–345.

Kodell, R.L., D. Krewski, and J.M. Zielinski. 1991. Additive and multiplicative relative risk in the two-stage clonal expansion model of carcinogenesis. Risk Anal. 11:483–490.

Konstantinova, I.V., E.N. Antropova, V.I. Legen'kov, and V.D. Zazhirei. 1973. Reactivity of lymphoid blood cells in the crew of Soyuz-6, 7, and 8 spacecraft before and after flight [in Russian]. Kosm. Biol. Med. 7:35–40.

Krewski, D., and D.J. Murdoch. 1990. Cancer modeling with intermittent exposures. Pp. 196–214 in Scientific Issues in Cancer Risk Assessment, S.H. Moolgavkar, ed. Boston: Birkhauser.

Krewski, D., and R.D. Thomas. 1992. Carcinogenic mixtures. Risk Anal. 12:105–113.

Krewski, D., D.J. Murdoch, and J. Withey. 1989a. Recent developments in carcinogenic risk assessment. Health Phys. 57(Suppl. 1):313–325.

Krewski, D., T. Thorslund, and J. Withey. 1989b. Carcinogenic risk assessment of complex mixtures. Toxicol. Ind. Health 5:851–867.

Leban, M.I., and P. A. Wagner. 1989. Space Station Freedom Gaseous Trace Contaminant Load Model Development. SAE Technical Paper Series 891513. Warrendale, Pa.: Society of Automotive Engineers.

Lehman, A.J., and O.G. Fitzhugh. 1954. 100-fold margin of safety. Assoc. Food Drug Off. U.S. Q. Bull. 18:33–35.

Lockhart, L.H. 1974. Cytogenetic studies of blood (experiment M111). Pp. 455–465 in Proceedings of the Skylab Life Sciences Symposium, August 27–29, 1974, Vol. 2. NASA TM X-58154. Houston, Tex.: Lyndon B. Johnson Space Center, National Aeronautics and Space Administration.

Lockhart, L.H. 1977. Cytogenetic studies of blood (experiment M111). Pp. 217–220 in Biomedical Results from Skylab, R.S. Johnston and L.F. Dietlein, eds. Washington, D.C.: National Aeronautics and Space Administration.

Lorenzi, G., P. Fuchs-Bislin, and A. Cogoli. 1986. Effects of hypergravity on "whole blood" cultures of human lymphocytes. Aviat. Space Environ. Med. 57:1131–1135.

Lutwak, L., G.D. Whedon, P.A. LaChance, J.M. Reid, and H.S. Lipscomb. 1969. Mineral, electrolyte and nitrogen balance studies of the Gemini-VII fourteen-day orbital space flight. J. Clin. Endocrinol. Metab. 29:1140–1156.

Lutz, W.K. 1990. Dose-response relationship and low-dose extrapolation in chemical carcinogenesis. Carcinogenesis 11:1243–1247.

Maigetter, R.Z., R. Ehrlich, J.D. Fenters, and D.E. Gardner. 1976. Potentiating effects of manganese dioxide on experimental respiratory infections. Environ. Res. 11:386–391.

Mandel, A.D., and E. Balish. 1977. Effect of space flight on cell-mediated immunity. Aviat. Space Environ. Med. 48:1051–1057.

Mantel, N., and W.R. Bryan. 1961. "Safety" testing of carcinogenic agents. J. Natl. Cancer Inst. 27:455–470.

Mantel, N., and M.A. Schneiderman. 1975. Estimating safe levels, a hazardous undertaking. Cancer Res. 35:1379–1386.

McColl, R.S. 1990. Biological Safety Factors in Toxicological Risk Assessment. SSC H49-49/1990E. Ottawa, Ont.: Supply and Services Canada. Available from NTIS, Springfield, Va., Doc. No. MIC90-06409.

McKusick, V.A. 1990. Mendelian Inheritance in Man. Catalogs of Autosomal Dominant, Autosomal Recessive and X-linked Phenotypes, 9th Ed. Baltimore, Md.: Johns Hopkins University Press.

McMillan, D.E. 1987. Risk assessment for neurobehavioral toxicity. Environ. Health Perspect. 76:155-161.

McMillan, D.E., and J.D. Leander. 1976. Effects of drugs on schedule-controlled behavior. Pp. 85-139 in Behavioral Pharmacology, S.D. Glick and J. Goldfarb, eds. St. Louis, Mo.: C.V. Mosby.

Melnick, R.L. 1984. Toxicities of ethylene glycol and ethylene glycol monoethyl ether in Fischer 344/N rats and B6C3F$_1$ mice. Environ. Health Perspect. 57:147-155.

Merrill, A.H., Jr., E. Wang, D.P. Jones, and J.L. Hargrove. 1987. Hepatic function in rats after spaceflight: Effects on lipids, glycogen, and enzymes. Am. J. Physiol. 252(2, Part 2):R222-R226.

Merrill, A.H., Jr., M. Hoel, E. Wang, R.E. Mullins, J.L. Hargrove, D.P. Jones, and I.A. Popova. 1990. Altered carbohydrate, lipid, and xenobiotic metabolism by liver from rats flown on Cosmos 1887. FASEB J. 4:95-100; erratum 4:2539.

Moolgavkar, S.H., and A.G. Knudson, Jr. 1981. Mutation and cancer: A model for human carcinogenesis. J. Natl. Cancer Inst. 66:1037-1052.

Moolgavkar, S.H., and G. Luebeck. 1990. Two-event model for carcinogenesis: Biological, mathematical, and statistical considerations. Risk Anal. 10:323-341.

Moolgavkar, S.H., and D.J. Venzon. 1979. Two-event models for carcinogenesis: Incidence curves for childhood and adult tumors. Math. Biosci. 47:55-77.

Munro, I.C., and D.R. Krewski. 1981. Risk assessment and regulatory decision making. Food Cosmet. Toxicol. 19:549-560.

Murdoch, D.J., and D. Krewski. 1988. Carcinogenic risk assessment with time-dependent exposure patterns. Risk Anal. 8:521-530.

Murdoch, D.J., D. Krewski, and J. Wargo. In press. Cancer risk assessment with intermittent exposure. In Proceedings of the Conference on Risk Assessment in the U.S. Department of Defense: Science, Policy, and Practice, H. Clewell, ed. Cincinnati, Ohio: American Conference of Government Industrial Hygienists.

NASA (National Aeronautics and Space Administration). 1988. Airborne Particulate Matter in Spacecraft. Proceedings of a Panel Discussion held at the Lunar and Planetary Institute, Houston Tex., July 23-24, 1987. NASA Conf. Publ. 2499. Houston, Tex.: Lyndon

B. Johnson Space Center, National Aeronautics and Space Administration.

NASA (National Aeronautics and Space Administration). 1989. Spacelab Life Sciences 1. First Space Laboratory Dedicated to Life Sciences Research. NASA Publ. NP-120. Houston, Tex.: Lyndon B. Johnson Space Center, National Aeronautics and Space Administration.

NASA (National Aeronautics and Space Administration). 1991. Flammability, Odor, Offgassing, and Compatibility Requirements and Test Procedures for Materials in Environments That Support Combustion. NHB 8060.1C. Houston, Tex.: Office of Safety and Mission Quality, Lyndon B. Johnson Space Center, National Aeronautics and Space Administration.

NRC (National Research Council). 1968. Atmospheric Contaminants in Spacecraft. Washington, D.C.: National Academy of Sciences.

NRC (National Research Council). 1970. Evaluating the Safety of Food Chemicals. Report of the Food Protection Committee. Washington, D.C.: National Academy of Sciences.

NRC (National Research Council). 1972. Atmospheric Contaminants in Manned Spacecraft. Washington, D.C.: National Academy of Sciences.

NRC (National Research Council). 1977. Drinking Water and Health. Washington, D.C.: National Academy Press.

NRC (National Research Council). 1979. Cardiovascular research. Pp. 27–42 in Life Beyond the Earth's Environment: The Biology of Living Organisms in Space. Washington, D.C.: National Academy Press.

NRC (National Research Council). 1984a. Emergency and Continuous Exposure Limits for Selected Airborne Contaminants, Vol. 1. Washington, D.C.: National Academy Press.

NRC (National Research Council). 1984b. Emergency and Continuous Exposure Limits for Selected Airborne Contaminants, Vol. 2. Washington, D.C.: National Academy Press.

NRC (National Research Council). 1984c. Emergency and Continuous Exposure Limits for Selected Airborne Contaminants, Vol. 3. Washington, D.C.: National Academy Press.

NRC (National Research Council). 1984d. Toxicity Testing: Strategies to Determine Needs and Priorities. Washington, D.C.: National Academy Press.

NRC (National Research Council). 1985a. Emergency and Continuous Exposure Guidance Levels for Selected Airborne Contaminants, Vol. 4. Washington, D.C.: National Academy Press.

NRC (National Research Council). 1985b. Emergency and Continuous Exposure Guidance Levels for Selected Airborne Contaminants, Vol. 5. Washington, D.C.: National Academy Press.

NRC (National Research Council). 1986a. Emergency and Continuous Exposure Guidance Levels for Selected Airborne Contaminants, Vol. 6. Washington, D.C.: National Academy Press.

NRC (National Research Council). 1986b. Criteria and Methods for Preparing Emergency Exposure Guidance Level (EEGL), Short-Term Public Emergency Guidance Level (SPEGL), and Continuous Exposure Guidance Level (CEGL) Documents. Washington, D.C.: National Academy Press.

NRC (National Research Council). 1987a. Emergency and Continuous Exposure Guidance Levels for Selected Airborne Contaminants, Vol. 7. Washington, D.C.: National Academy Press.

NRC (National Research Council). 1987b. The application of pharmacokinetic data in carcinogenic risk assessment. Pp. 441-468 in Pharmacokinetics in Risk Assessment, Drinking Water and Health, Vol. 8. Washington, D.C.: National Academy Press.

NRC (National Research Council). 1987c. Allometry: Body size constraints in animal design. Pp. 65-79 in Pharmacokinetics in Risk Assessment, Drinking Water and Health, Vol. 8. Washington, D.C.: National Academy Press.

NRC (National Research Council). 1988a. Emergency and Continuous Exposure Guidance Levels for Selected Airborne Contaminants, Vol. 8. Washington, D.C.: National Academy Press.

NRC (National Research Council). 1988b. Space Science in the Twenty-first Century: Imperatives for the Decades 1995 to 2015. Life Sciences. Washington, D.C.: National Academy Press.

NRC (National Research Council). 1988c. Complex Mixtures. Methods for In Vivo Toxicity Testing. Washington, D.C.: National Academy Press.

NRC (National Research Council). 1989a. Recommended Dietary Allowances, 10th Ed. Washington, D.C.: National Academy Press.

NRC (National Research Council). 1989b. Biologic Markers in Pulmonary Toxicology. Washington, D.C.: National Academy Press.

OSTP (Office of Science and Technology Policy). 1985. Chemical carcinogens: A review of the science and its associated principles. Fed. Regist. 50:10371-10442.

Travis, C.C., and R.K. White. 1988. Interspecific scaling of toxicity data. Risk Anal. 8:119–125.

Wallace, L.A. 1987. The Total Exposure Assessment Methodolgoy (TEAM) Study: Summary and Analysis, Vol. I. EPA 600/6-87/002a. Washington, D.C.: U.S. Environmental Protection Agency.

Waxweiler, R.J., W. Stringer, J.K. Wagoner, J. Jones, H. Falk, and C. Carter. 1976. Neoplastic risk among workers exposed to vinyl chloride. Ann. N.Y. Acad. Sci. 271:40–48.

Weinberg, R.A. 1991. Tumor suppressor genes. Science 254:1138–1146.

Whedon, G.D. 1984. Disuse osteoporosis: Physiological aspects. Calcif. Tissue Int. 36:S146–S150.

Whedon, G.D., L. Lutwak, J. Reid, P. Rambaut, M. Whittle, M. Smith, and C. Leach. 1974. Mineral and nitrogen metabolic studies on Skylab orbital space flights. Trans. Assoc. Am. Physicians 87:95–110.

Whittemore, A., and J. Keller. 1978. Quantitative theories of carcinogenesis. SIAM Rev. 20:1–30.

Whittemore, A.S., and A. McMillan. 1983. Lung cancer mortality among U.S. uranium miners: A reappraisal. J. Natl. Cancer Inst. 71:489–499.

Wyrobek, A.J., and W.R. Bruce. 1978. The induction of sperm-shape abnormalities in mice and humans. Pp. 257–285 in Chemical Mutagens: Principles and Methods for Their Detection, Vol. 5, A. Hollaender and F.J. de Serres, eds. New York: Plenum.

Wyrobek, A.J., M. Currie, J.L. Stilwell, R. Balhorn, and L.H. Stanker. 1990. Detecting specific-locus mutations in human sperm. Pp. 93–109 in Biology of Mammalian Germ Cell Mutagenesis. Banbury Report 34, J.W. Allen, B.A. Bridges, M.F. Lyon, M.J. Moses, and L.B. Russell, eds. Cold Spring Harbor, N.Y.: Cold Spring Harbor Laboratory Press.

Appendix 1

FORMAT FOR SMAC DOCUMENTS

Background Information
  Physical and Chemical Properties
  Occurrence and Use

Summary of Toxicity Information
  Effects on Humans
  Effects on Animals
    Acute, Subacute, and Chronic Exposures
    Mutagenicity, Teratogenicity, and Carcinogenicity

Pharmacokinetics
  Absorption and Distribution
  Metabolism and Excretion

Inhalation Exposure Levels (from other sources, such as TLVs from ACGIH and PELs from OSHA standards)

Committee Recommendations (emergency exposure guidance levels: current and prior COT recommendations and rationale for new numbers)

Recommendations for Future Research (when applicable)

Tables

References (and cutoff date for published papers)

Appendix 2

CONVERSION FACTORS

1. From gas in gas to ppm by volume, at 25°C and 760 mmHg:

$$\frac{mg}{m^3} \times 10^{-3} = \frac{mg}{L}$$

$$\frac{mg}{m^3} \times \frac{24{,}450}{(molecular\ wt)} = ppm$$

$$\frac{\mu m\ of\ gas}{mol\ of\ air} = ppm$$

$$\%\ by\ volume \times 10^{-4} = ppm$$

2. From gas, liquid, or solid in liquid to ppm by weight:

$$\frac{mg}{L} = ppm$$

$$\frac{mol}{L} \times molecular\ wt \times 10^3 = \frac{mg}{L}$$

3. From amount received by inhalation to amount ingested:

$$\frac{mg}{m^3} \times volume\ of\ inspired\ in\ air\ m^3 \times \%\ retention/100 = mg$$

4. From concentration in diet to ingested dose:

$$ppb \text{ in diet} \times 10^3 = ppm \text{ in diet}$$

$$ppm \text{ in diet} = mg/kg \text{ in diet} = \mu g/g \text{ in diet}$$

$$mg/kg \text{ diet} \times \frac{food\ intake,\ kg/day}{body\ weight,\ kg} = mg/kg/day$$

5. From percent to concentration:

$$wt/wt\ \% = 10^{-2}\ kg/kg$$
$$vol/vol\ \% = 10^{-2}\ L/L$$
$$wt/vol\ \% = 10^{-2}\ kg/L$$
$$g\ \% = g/100\ ml$$

Appendix 3

REFERENCE VALUES USED BY COT

	Man	Woman
Daily water intake, adult human	2 L	
Daily food intake, rat	20 g	
Body Weight, kg	70	58
Blood volume, L	5.2	3.9
Total blood weight, g	5,500	4,100
Red-cell volume, L	2.2	1.35
Red-cell weight, g	2,400	1,500
Plasma volume, L	3.0	2.5
Plasma weight, g	3,100	2,600
Urine volume, L/day	1.4	1.0
Surface area, cm^2	18,000	16,000
Minute volume, resting, L/min	7.5	6.0
Minute volume, light activity, L/min	20.0	19.0
Volume inspired air, 8-hr workday, m^3	9.6	9.1
Volume inspired air, rest all day, m^3	15	
Volume inspired air, moderate activity plus sleep, m^3	20	

www.ingramcontent.com/pod-product-compliance
Lightning Source LLC
Chambersburg PA
CBHW081727170526
45167CB00009B/3733